Lecture Notes in Electrical Engineering

Volume 508

**** Indexing: The books of this series are submitted to ISI Proceedings, EI-Compendex, SCOPUS, MetaPress, Springerlink ****

Lecture Notes in Electrical Engineering (LNEE) is a book series which reports the latest research and developments in Electrical Engineering, namely:

- Communication, Networks, and Information Theory
- Computer Engineering
- Signal, Image, Speech and Information Processing
- Circuits and Systems
- Bioengineering
- Engineering

The audience for the books in LNEE consists of advanced level students, researchers, and industry professionals working at the forefront of their fields. Much like Springer's other Lecture Notes series, LNEE will be distributed through Springer's print and electronic publishing channels.

For general information about this series, comments or suggestions, please use the contact address under "service for this series".

To submit a proposal or request further information, please contact the appropriate Springer Publishing Editors:

Asia:

China, *Jessie Guo, Assistant Editor* (jessie.guo@springer.com) (Engineering)

India, *Swati Meherishi, Senior Editor* (swati.meherishi@springer.com) (Engineering)

Japan, *Takeyuki Yonezawa, Editorial Director* (takeyuki.yonezawa@springer.com) (Physical Sciences & Engineering)

South Korea, *Smith (Ahram) Chae, Associate Editor* (smith.chae@springer.com) (Physical Sciences & Engineering)

Southeast Asia, *Ramesh Premnath, Editor* (ramesh.premnath@springer.com) (Electrical Engineering)

South Asia, *Aninda Bose, Editor* (aninda.bose@springer.com) (Electrical Engineering)

Europe:

Leontina Di Cecco, Editor (Leontina.dicecco@springer.com)

(Applied Sciences and Engineering; Bio-Inspired Robotics, Medical Robotics, Bioengineering; Computational Methods & Models in Science, Medicine and Technology; Soft Computing; Philosophy of Modern Science and Technologies; Mechanical Engineering; Ocean and Naval Engineering; Water Management & Technology)

Christoph Baumann (christoph.baumann@springer.com)

(Heat and Mass Transfer, Signal Processing and Telecommunications, and Solid and Fluid Mechanics, and Engineering Materials)

North America:

Michael Luby, Editor (michael.luby@springer.com) (Mechanics; Materials)

More information about this series at http://www.springer.com/series/7818

Sri Niwas Singh · Fushuan Wen
Monika Jain
Editors

Advances in Energy and Power Systems

Select Proceedings of ICAEDC 2017

Springer

Editors
Sri Niwas Singh
Madan Mohan Malaviya University of Technology
Gorakhpur, Uttar Pradesh
India

Fushuan Wen
Department of Electrical Engineering
Zhejiang University
Hangzhou, Zhejiang
China

Monika Jain
Department of Electrical and Electronics Engineering
I.T.S. Engineering College
Greater Noida, Uttar Pradesh
India

ISSN 1876-1100 ISSN 1876-1119 (electronic)
Lecture Notes in Electrical Engineering
ISBN 978-981-13-0661-7 ISBN 978-981-13-0662-4 (eBook)
https://doi.org/10.1007/978-981-13-0662-4

Library of Congress Control Number: 2018941992

Printed on acid-free paper

This Springer imprint is published by the registered company Springer Nature Singapore Pte Ltd.
part of Springer Nature
The registered company address is: 152 Beach Road, #21-01/04 Gateway East, Singapore 189721, Singapore

Preface

This volume contains 18 papers from International Conference on Advancement in Energy, Drive & Control (ICAEDC 2017), which is devoted to the gamut of emerging engineering issues from theoretical aspects to application-dependent studies and the validation for a specific application. This was envisioned and founded to represent the growing needs of research in the areas of electrical, computer and electronics engineering and technology. This proceeding is intended as a forum for practitioners and researchers to share their ideas in the field of sustainable energy, intelligent control, signal processing and communication.

The format consists of plenary sessions, tutorials and workshops on a niche area. Around 300 people have participated in ICAEDC-2017, 186 papers have been received, and approximately selected 90 papers have been presented by authors.

The conference was inaugurated by honourable Chief Guest Mr. Dinesh Jain, President, Legal and Corporate Affairs, Uflex Ltd., India, followed by a first plenary session from distinguished speaker Prof. Abul Hassan Siddiqi, Ex Pro Vice Chancellor, AMU, and Visiting Consultant, ICTP, entitled "Mathematical Models & Methods of Oil Industry". In his session, some well-known mathematical models in the form of a partial differential equation representing the real-world system and their wavelet-based simulation were presented. The plenary session on "Building a Cyber Resilient Energy Sector" was delivered by Dr. Anuj Goel, SM-IEEE, Co-Founder and CEO, Cyware Labs, USA. Further, keynote speaker Prof. Prerna Gaur, SM-IEEE, NSIT, New Delhi, has delivered the session on "Application of Doubly Fed Induction Generator and Permanent Magnet Synchronous Generator for Variable Speed Wind Turbines". The next session was on "Applications of Mathematical Modeling in Energy Drives" which is considered as a standard tool in energy planning and management, delivered by Prof. V. N. Jha, Prince Sattam Bin Abdulaziz University, Kingdom of Saudi Arabia. The last keynote session was by Prof. Seema Arora, Waljat College of Applied Sciences, Muscat, Oman, on "Broadband Over Power Lines and Smart Grids" which defines integration of existing electrical infrastructure with the communication infrastructure that allows power and bidirectional information flow efficiently.

The conference was also featured with four tutorial sessions. The first tutorial was delivered by Mr. Antonio VILEI, Software Platforms and Cloud, STMicroelectronics, Lecce, Italy, on "The Internet of Things Revolution: Smart Cities and Smart Industry". In parallel to this, the second tutorial was on "Impacts of Artificial Intelligence and Machine Learning on Workforce and Society, a Macro Economic Perspective" by Mr. Anshuman Tripathi, Head Consultant, Starburst, USA. Mr. Anil Kumar Goel, Ex DGM, NTPC, has delivered the third tutorial on "Power Scenario in India and Present Challenges". Professor M. A. Ansari, SM-IEEE, Gautam Buddha University, has delivered the fourth tutorial session on "Renewable Energy Scenario and Future Scope in India".

The workshop was another feature of this conference. It was scheduled in successive sessions on "Trends in Microgrid Management" by Mr. Himadri Endow, Ex Head Smart Grid, Alstom Grid, India–Southeast Asia, followed by another workshop on "Advancement and Skill Challenges in Instrumentation and Automation" by Mrs. R. Mahalakshmi, DGM, Yokogawa India Limited. Mr. Piyush Chandra Ojha, VP CABCON, India, has delivered the third workshop on "Microgrids—The Changing Paradigm for Grid Control". The fourth workshop on "IOT in Building and Home Automation" was delivered by Mr. Nitin Jain, Director and Chief Operating Officer, Aviconn Solutions Pvt. Ltd. New opportunities of Automation, Optimisation and Monetisation are being identified everyday in real estate segments with the virtue of IoT, Artificial Intelligence (AI) and Data Analytics.

The ICAEDC 2017 has received the partial financial support from Science and Engineering Research Board (SERB), Department of Science and Technology, Government of India. The conference has also received its technical support from IEEE UP section.

Different chapters of this volume are categorised into the following major technologies:

- Economical Assessment of Electric Utilities,
- Sustainable Energy and Performance Analysis,
- Power Quality Enhancement,
- Power Line Communication,
- Intelligent Control Systems,
- Reliability Measures and Optimization,
- Antenna Design and Processing,
- Wireless, Private Security and Routing Techniques,
- Biomedical and Biometric Applications,
- Artificial Intelligence and Optimization and
- Signal and Image Processing.

Economical Assessment of Electric Utilities

Since the operating characteristics of a modern power system are becoming increasingly complicated, it is imperative to assess the daily power system dispatching quality. Hence, Pengcheng Cui et al. have presented a post-evaluation index system for daily power system dispatching, based on the average effect, buckets effect and abnormalities effect, from the perspectives of system security, economics, energy saving, environmental friendliness and impartiality. Stated case studies show that the proposed method can not only identify weak links, but also reflect the overall level accurately and is feasible and effective for applications in practical power systems. Kalyan Chaterjee et al. have presented a technique for calculating comparative cost efficiencies of Indian State Owned Electric Utilities (SOEUs), which have been primarily accountable for the distribution of electricity in India. Performance of 28 state utilities was evaluated using the nonparametric approach of data envelopment analysis (DEA).

Sustainable Energy and Performance Analysis

As renewable energy has got a forefront focus for power generation due to the gradual depletion of fossil fuel and increased concerns about global warming, challenges associated with wind power plant has also become a major concern for researchers. As wind speed, being stochastic in nature, is making energy extracted from it nondispatchable, hence Singh et al. have proposed a method to regulate it by coupling the wind energy system (WES) with flywheel, an energy storage device which requires less maintenance and is quick acting, mechanically robust, high energy efficient, least environment pollutant. It quickly reacts to variation in the electrical output of wind energy system, thereby delivering smoother power to grid. Active power regulation due to 500 kW, 100 kgm^2 flywheel energy system integrated with 1.5-MW type-3 wind turbine generator is considered for variable wind condition in MATLAB/Simulink. Currently, solar energy is identified as a sustainable renewable energy source which is catching the eyes of big power industries and entrepreneurs who are constantly doing research for enhancing efficiency and cost of energy production. Towards those aspects, the authors have proposed a performance analysis of PV cell using one-diode and two-diode models. Another author has presented a review work of various maximum power point tracking (MPPT) techniques which experience partial shading conditions (PSCs), an unavoidable complication that significantly reduces the efficiency of the overall system. The exhaustive comparison of various MPPTs taken by the researchers which tracks the global peak (GP) of a photovoltaic (PV) array under PSC is made, and the best available option is proposed for researchers. MPPT techniques such as perturb and observe (P&O), improved particle swarm optimization (IPSO) and grey wolf optimization (GWO) are the recent techniques proposed by the researchers.

Various DC–DC converter topologies with PV system are also compared for the purpose of maximum power point tracking.

Power Quality Enhancement

The penetration of renewable energy is increasing worldwide. The advancement in renewable energy is exciting, but at the same time it is creating significant technical challenges to the power industry. The term "power quality" is a broad concept. It is associated with electrical transmission, distribution and utilization systems. Due to power quality problems, industries have to invest large amount for mitigation of voltage sags, distortions, harmonics and short-term interruptions/disturbances. Hence, authors have proposed to cover various possible sources and compensation methods of reactive power in power system under different contingencies using FACTS devices and advanced power converters.

Power Line Communication

Distribution load management is an important concern for the utilities to exploit the existing infrastructure of the power system network. There is a need to improve the reliability of power line networks and provide standards and systems for utilizing power networks as a medium of communication. There are areas where the digital communication may be made through low-voltage distribution network. This helps the utilities to reach remote locations. Mapping of the consumer through control centre is achieved through interfacing of coupling circuits that leads to distribution load management. Arora and Jain present designing of broadband coupling circuit that satisfies specific signal transmission, appropriate bandwidth and limited number of components. The paper also discusses the significant parameter variations during capacitive couplings and inductive couplings for passive and active network topologies.

Intelligent Control Systems

To avoid a catastrophic failure in any process plant, precise control of all associated drives is critical for real-time application. Singh et al. have proposed to implementation of ANFIS controller-based algorithm in precise control techniques particularly suitable in the textile, aerospace and automobile sectors. In this chapter, a method has been proposed by implementing artificial intelligence (AI)-based software algorithm which tracks the position of the motor from signals received from the resolver and thereby calculates speed from the tracked position. Further, they have presented a comparison between AI-based controller and PI controller showing how it eliminates the delay and noise effects introduced by hardware circuit and can be readily implemented in real time with minimum complexity and cost. In a subsequent chapter, for the complex higher-order system analysis, several order reduction techniques have been proposed. In another paper, a mixed method is proposed which combines the improved Pade approximations and the eigen spectrum analysis for reducing the higher-order system. In this method, system stiffness and pole centroid of both original and reduced-order systems remain same. The denominator of higher-order system (HOS) is derived using the eigen spectrum analysis, and the numerator is derived by using improved Pade approximation. The later method of order reduction utilizes both time moments and Markov parameters. The stability and quality of the reduced-order system are compared with the existing methods of order reduction. To understand the proposed method, the paper includes some numerical examples of single-input single-output (SISO) systems.

Reliability Measures and Optimization

As complexity increases the system unavailability, reliability analysis plays an important role for its assessment to ensure the performance and availability of any specific systems at the point of requirement. Most preferred techniques of reliability optimization include the incorporation of redundant element at either component or system level, and generally, the redundant components are of identical types. But the occurrence of common cause failure (CCF) defeats the benefit of redundancy optimization. To validate the same, authors have presented five exact approaches to evaluate the reliability of the complex network and also extended the work to evaluate and compare the reliability of the same network under CCF. To incorporate the CCFs in reliability evaluation, conditional probability approach is used.

Antenna Design and Processing

Articles here exemplify the analysis and exploration of communication systems. They provide invaluable insights into the improvement of bandwidth and efficiency of types of antennas used for communication systems, the wide-band antenna for millimetre-wave application, where a single antenna is used rather than more antennas in a device, to reduce the size and cost of the device. An investigation into the design of Giuseppe Peano fractal patch antenna and reactive impedance surface (RIS) is discussed for the improvement of bandwidth and improvement in antenna radiation. A review of metamaterial structures is also presented for the enhancement of various antenna properties. The objective of these articles is to get the best configuration required for the desired specification and also helps in developing the future ideas by taking into account the advantages of the available structures.

Wireless, Private Security and Routing Techniques

In this section, the chapter on spectrum sensing technique for cognitive radio network and enhancement of trunking efficiency of mobile switching centre (MSC) is presented. Further, privacy issues of user identification by the use of encryption for pseudonymization using MAC and HMAC have also been illustrated.

Biomedical and Biometric Applications

Authors have focused on aspects related to EMD, which is applied to EEG recordings from epileptic patients. The classification was done by using these selected features by artificial neural network (ANN). Next articles illustrate signal processing in the field of biometrics. In the next chapter, low-power EEG-based brain–computer interface (BCI) for the physically challenged section of the society has been presented.

Artificial Intelligence and Optimization

In the very first article, authors have demonstrated the importance of integrated novel image processing techniques—which shows the potential of such intelligent approaches. Hence, to illustrate the application, an intelligent approach based on generalized neural network (GNN) is proposed and applied for the short-term solar PV power forecasting. In the next chapter, Volterra filtering, for feed-forward active noise controls (ANC), suitable for the nonlinear controller has been illustrated. The use of soft computing methods, especially neural networks, for filterring of noise from an image is proposed and presented. Noises can occur in images during acquisition on transmission. The presence of noise can hinder the proper utilization of these images for various applications such as medical imaging, satellite imaging. The use of soft computing methods, especially neural networks, is reviewed in-depth in

Signal Processing and Advanced Digital Design

The papers selected to be included in this category contribute to the understanding of relevant trends of current research on signal and image processing technologies. With context to that, a focus on a recursive algorithm for the computation of discrete cosine transform (DCT) and inverse discrete cosine transform (IDCT) has been proposed; algorithm is both time and hardware efficient. In the next article, wavelet shrinkage function (WSF) used for fault detection and diagnosis in a transmission line has been presented. In a subsequent chapter, field-programmable gate array (FPGA)-based architecture for implementation of DST has been proposed. To establish the communication between the blocks operating in different clock domains is challenging task due to synchronization failure. Therefore, in the next article, mixed clock first in, first out (FIFO) as an interface between different clock domains is briefed.

We immensely thank all the participants and authors for their contributions to ICAEDC 2017. The review committee has done an excellent job in reviewing the articles and approving the high-quality research articles to be published in this conference proceedings. The editors are thankful to various committees for their dedication in making this a very successful conference. We are grateful to Springer for making possible the publication of these proceedings of ICAEDC 2017. We would like to express our heartfelt thanks to college administration for their constant encouragement and facilitating with all logistics support. A special thanks go to core committee members Ms. Kalpana Hazarika and Mr. Nitin Kathuria, Assistant Professors, EEE Department, I.T.S. Engineering College, Greater Noida. Also, a deep sense of appreciation is extended to entire ICAEDC organizing committee for this successful endeavour.

Sincere gratitude is extended to Honorary Advisors of ICAEDC 2017: senior advisors: Prof. Abul Hassan Siddiqi, Ex Pro Vice Chancellor, AMU, and Visiting Consultant, ICTP, and Prof. Krishna B. Misra, Founder and Past Editor-in-Chief, *International Journal of Performability Engineering*, Editor, Book Series on *Performability Engineering*. We are also pleased to thank Dr. Manoj Kumar—Senior Group Manager, R&D and Applications Group at STMicroelectronics, India.

We sincerely hope that this volume will inspire researchers and scholars for innovation.

Sri Niwas Singh
Fushuan Wen
Monika Jain
Editors of ICAEDC 2017

Contents

Part I Renewable Energy

Maximum Power Point Tracking Approaches for Wind–Solar Hybrid Renewable Energy System—A Review 3
Surabhi Chandra, Prerna Gaur and Srishti

Short-Term PV Power Forecasting Using Generalized Neural Network and Weather-Type Classification 13
Priyanka Chaudhary and M. Rizwan

Model Order Reduction of Two Area Hydropower System Using Mixed Method 21
Sadaf Naqvi, Ibraheem Nasiruddin, Sana Ali and Shilpa Gupta

Current Trends in Control Techniques in Renewable Energy: A Review 31
Srishti, Prerna Gaur and Surabhi Chandra

Performance Analysis of PV Cell Using One Diode and Two Diode Model 43
Noorul Islam, A. Q. Ansari and L. Navinkumar Rao

Part II Power Management

Active Power Regulation by MPC based Flywheel Energy Storage System 57
Aasim, S. N. Singh and Abheejeet Mohapatra

Post-evaluation Index System and Comprehensive Evaluation Method for Daily Power System Dispatching 73
Pengcheng Cui, Bing Xu, Fushuan Wen, Zihao Fu, Bingquan Zhu, Qiulong Ni and S. N. Singh

Real Assessment of State Owned Indian Electric Utilities 87
R. V. S. Sengar, Kalyan Chattterjee and Jay Singh

A Novel Scheme of Fault Detection in Transmission Line Using Image Processing 103
Deepak Kumar, Amit Kumar, Abhay Yadav and M. A. Ansari

Power Quality Improvement of Power Distribution System Under Symmetrical and Unsymmetrical Faults Using D-STATCOM 111
Amit Kumar, Deepak Kumar and Abhay Yadav

Optimal Sitting and Sizing of Capacitor Using Iterative Search Method for Enhancement of Reliability of Distribution System 123
Vikas Singh Bhadoria, Nidhi Singh Pal, Vivek Shrivastava and Shiva Pujan Jaiswal

Battery Energy Storage Technology Integrated for Power System Reliability Improvement 131
Same Ram Ramavat, Shiva Pujan Jaiswal, Nitin Goel and Vivek Shrivastava

Thermal Evaluation and Oxidation Stability of High Temperature Alternative Solid Dielectrics of Power Transformers in Mixed Oil 141
Manoj Kumar, Chilaka Ranga, Ashwani Kumar Chandel and Ashish Kumar Mishra

Multi-area Economic Dispatch Using Dynamically Controlled Particle Swarm Optimization 151
Vineet Kumar, Ram Naresh Sharma and Shiv Kumar Sikarwar

An Analytical Hierarchy Process Based Approach for Effective Maintenance Prioritization of Power Transformers 165
Kavish Jain, Awin Gupta, Yog Raj Sood and Manisha Sharma

Designing of Controllers Using Chebyshev–Pole Clustering Approximants 175
Vartika Rao

Comparison: Matrix Converter, Cycloconverter, and DC Link Converter 185
Gurmeet Kaur and Sandeep Kumar

Power Quality Improvement and Analysis Using Multi-pulse Converter 193
Renu Kumari, M. A. Ansari and Shubham Shukla

About the Editors

Sri Niwas Singh is currently the vice chancellor of Madan Mohan Malviya University of Technology, Gorakhpur, India, and a professor at the Department of Electrical Engineering, Indian Institute of Technology Kanpur, India. He obtained his M.Tech. and Ph.D. in electrical engineering from the Indian Institute of Technology Kanpur in 1989 and 1995, respectively. He was an assistant engineer at UP State Electricity Board from 1988 to 1996 and was an assistant professor at Roorkee University (now IIT Roorkee) from 1996 to 2000 and at the Asian Institute of Technology, Bangkok, Thailand, from 2001 to 2002. He has received numerous awards, including Young Engineer Award from the Indian National Academy of Engineering (INAE), Khosla Research Award of IIT Roorkee and Young Engineer Award of CBIP New Delhi (India). He is also the recipient of the German Humboldt Fellowship and the Danish Otto Monsted Fellowship. In 2013, he became the first Asian to receive IEEE Educational Activity Board Meritorious Achievement Award in continuing education. He was also the recipient of the INAE Outstanding Teacher Award in 2016 and IEEE R10 region (Asia-Pacific) Outstanding Volunteer Award in 2016. His research interests include power system restructuring, FACTS, power system optimization and control, wind power and security analysis. He has published more than 440 papers in international/national journals/conferences and written two books.

Fushuan Wen is a professor in the School of Electrical Engineering at Zhejiang University, China. He completed his B.E. and M.E. at Tianjin University, Tianjin, China, in 1985 and 1988, respectively, and went on to do his Ph.D. at Zhejiang University, Hangzhou, China, in 1991, all in electrical engineering. He joined the faculty of Zhejiang University in 1991 and has been a full professor and the director of the Institute of Power Economics and Information since 1997 and the director of Zhejiang University-Insigma Joint Research Center for Smart Grids since 2010. He has undertaken various teaching, research and visiting appointments at the National University of Singapore (NSTB Postdoctoral Fellowship, Research Fellowship), Hong Kong Polytechnic University (Research Fellowship, Visiting Assistant Professorship), University of Hong Kong (Research Assistant Professorship), South

China University of Technology (University Distinguished Professorship), University of New South Wales (ARC Senior Fellowship), Queensland University of Technology (CSIRO Visiting Fellowship), University of Technology Brunei (Professorship in Power Systems), Technical University of Denmark (Otto Monsted Guest Professorship in Power Systems), Nanyang Technological University (Visiting Fellowship). His research interests include power economics and electricity markets; power system investment; planning and operation optimization; smart grids and electric vehicles; power system alarm processing; fault diagnosis and system restoration. He is an editor of *IEEE Transactions on Power Systems* and *IEEE Power Engineering Letters* and associate editor of *IET Generation, Transmission and Distribution*, the *Journal of Energy Engineering* (ASCE) and the *Journal of Modern Power Systems and Clean Energy* (Springer). He is also on the editorial boards of more than ten journals.

Monika Jain is a professor and head in the Department of Electrical and Electronics Engineering at ITS Engineering College, Greater Noida. She is a senior administrator and passionate educationist with more than 20 years of experience in various prestigious institutions. She completed her M.Sc. in instrumentation at IIT Roorkee and subsequently an M.Tech. from DAVV Indore and Ph.D. from BITS, Pilani. She has been the recipient of many awards and travel grants and has convened various high-profile conferences and seminars sponsored by AICTE, DST, IEEE and other funding agencies/sponsors. She has received several financial grants from the Government of India for hosting these events. Her major contributions include the establishment of a "Project Lab on Virtual Instrumentation" funded by AICTE under the "MODROBS" scheme. She has published more than 40 research papers in respective international journals and conferences, and her Ph.D. work was published in *IEEE Transactions on Broadcasting*. Her research interests include MPEG systems, set-top box domains, embedded systems, IoTs, PLC, SCADA and automation. She serves as a reviewer for various leading conferences and journals. She is a member of IEEE, life member of IETE and member of the Indian Society of Industrial and Applied Mathematics (ISIAM).

Part I
Renewable Energy

Maximum Power Point Tracking Approaches for Wind–Solar Hybrid Renewable Energy System—A Review

Surabhi Chandra, Prerna Gaur and Srishti

Abstract For effective energy conversion from renewable energy resources, maximum power point tracking (MPPT) algorithms are gaining significance due to their effectiveness and adaptability to handle nonlinear conditions. This paper reviews influence of numerous real-time MPPT control methodologies for wind–solar hybrid renewable energy system (HRES). From the past research surveys, this paper will present the common trends in MPPT schemes and a discussion on advanced trends in intelligent techniques for wind–solar hybrid energy system. An evaluation on state of the art and performance of new intelligent control techniques on several criteria, i.e., array dependency, wind speed, convergence time, handling shading conditions, implementation for practical work, will also be elucidated. This review work is to foresee the application of advanced artificial intelligence (AI) control strategies to serve as a reference for new developments in powerful usage of hybrid energy system.

Keywords MPPT · Solar · Wind · Hybrid renewable energy system
Artificial intelligence

1 Introduction

Due to the rapid exhaustion of fossil fuels and their effects like environmental hazards, the power systems are now upgrading toward green power generation [1, 2]. The widespread increase of renewable energy sources permits attaining more safe and viable energy prospects to overcome the challenge of growing energy consumption and utilization. The trend of renewable energy power generation is now heading

S. Chandra (✉) · P. Gaur · Srishti
Netaji Subhas Institute of Technology, New Delhi, Delhi, India
e-mail: surabhichandra.en@gmail.com

P. Gaur
e-mail: prernagaur@yahoo.com

Srishti
e-mail: srishti_26@yahoo.com

S. N. Singh et al. (eds.), *Advances in Energy and Power Systems*, Lecture Notes in Electrical Engineering 508, https://doi.org/10.1007/978-981-13-0662-4_1

toward generation of power from hybrid renewable energy resources, i.e., solar energy and wind energy.

According to Ministry of New and Renewable Energy (MNRE), the total Installed Grid Interactive Renewable Power Capacity in India as of July 31, 2016, is 44,783.33 MW and by 2022 it is targeted to rise around 175,000.00 MW. The major contributors in renewable energy generations are wind and solar energy [3–5].

Wind and solar renewable energy resources are intermittent in nature. A system comprising of wind and solar combined together as hybrid system can be designed to increase the power generation [6]. As these two resources are naturally available in nature, so they are also dependent upon climatic conditions. Wind power output varies with wind speed, and solar power is dependent upon solar irradiance. Hence, the power output keeps fluctuating depending upon the resource available. Therefore to extract maximum power from the renewable energy resources, maximum power tracking algorithms are implemented.

There are variety of MPPT techniques to improve the efficiency of renewable energy systems like perturb and observe [7–9], incremental conductance [10], current sweep, constant voltage, distributed MPPT, hill climbing, slide control method, ripple correlation control, DC link capacitor droop control, gauss newton technique, adaptive P&O, curve fitting technique, parasitic capacitance technique, linearization-based MPPT technique, dP/dV or dP/dI feedback control. Nowadays, intelligent MPPT control techniques are opening broad ways in research fields. Fuzzy logic, neural networks, genetic algorithms, evolutionary algorithms, expert systems, swarm intelligence systems like ant colony system, particle swarm optimization, artificial bee colony system, cuckoo search algorithms are some of the new techniques which are evolving very fast for MPPT systems.

2 Hybrid Renewable Energy System

HRES incorporates different renewable energy resources. In this paper, HRES is formed by integrating solar and wind energy conversion system. Both systems are coupled to the DC bus. To obtain maximum power, MPPT control is applied at the converter. DC is converted to AC by inverter which supplies AC to the grid (Fig. 1).

3 Maximum Power Point Tracking

3.1 PV *System*

The purpose of using MPPT is to make sure that in environmental situation like solar irradiance and temperature changes *PV* modules are able to supply maximum power.

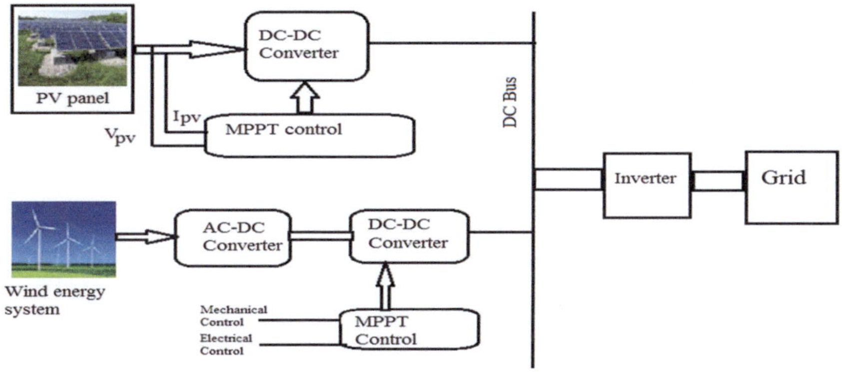

Fig. 1 Block diagram of HRES

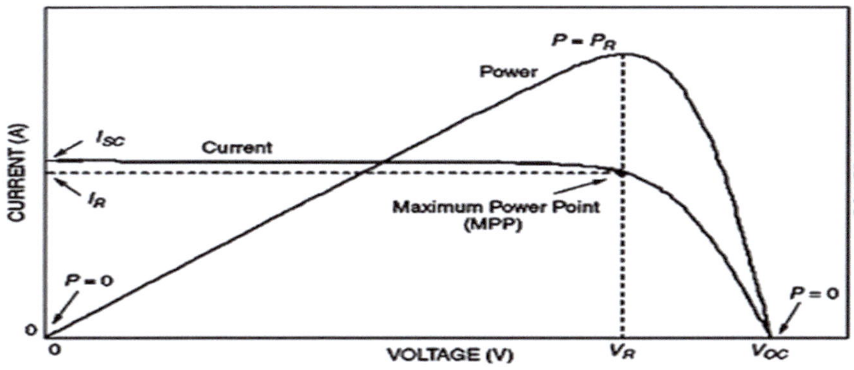

Fig. 2 MPPT characteristic of a *PV* cell

El-Khozondar [11] shows the typical maximum power point characteristic for a *PV* system. A characteristic curve has been plotted between voltage and current obtained from the *PV* cell, and the point of maxima of the curve is the point at which the *PV* cell should operate in order to generate maximum power (Fig. 2; Table 1).

3.2 Wind Energy System

The power output of wind turbine is proportional to the cubic function of the turbine speed. With the changes in wind speed, the wind energy systems are able to supply maximum power. Wei [12] has shown the wind turbine characteristic for wind turbine shaft speed to turbine power and has also obtained the optimal power curve at different wind speed. At the point where turbine power is maximum with respect to wind speed is maximum power point for wind turbine systems (Fig. 3).

Table 1 MPPT techniques for *PV* system

S. No.	MPPT technique	Remarks
1	Constant voltage	In low insulation environment, the constant voltage technique is more effective than P&O and IC method
2	Short current pulse method [15]	This method is adaptive to disturbances. By the help of proportional of parameter in a low-output-power region, MPPT performance is improved remarkably, especially in partial shading conditions and surface contamination
3	Open voltage method [16]	This method is based on constant voltage method, but it makes the assumption that the MPP voltage is always around 75% of the open-circuit voltage. So mainly, this technique takes into account the temperature. Besides, this technique can partially take into account the cell's aging
4	Perturb and observe method [17]	This technique commonly exhibits decent performance as long as the solar irradiation does not vary too quickly. The operating point usually fluctuates and oscillates around maximum power point
5	Incremental [17] conductance method	The benefit of this algorithm is that it is capable of tracking the maximum power point more specifically in different weather conditions and has the ability to determine the relative "distance" to the maximum power point (MPP)
6	Temperature gradient algorithm [18]	The temperature algorithm requires datasheet information regarding the *PV* array, and the algorithm has to be updated to ensure accurate operation of the *PV* system and compensate for parameter changes caused by system aging
7	I & T method [19]	The irradiance is estimated through the relation of the measured short-circuit current and its reference at standard test condition. This method uses low-cost sensors to measure temperature and current, which accurately define the MPP
8	Curve fitting technique	Offers simple indirect control of voltage *V* through a digital circuitry for a stand-alone system by modelling off-line based on mathematical equations or numerical approximations
9	Pilot cell algorithm [20]	This method is based on measuring short-circuit current and open-circuit voltage on a pilot cell (it is like a small solar cell having the characteristic of cells in the large array). This method leads to elimination of *PV* power loss during OC voltage and SC current measurement. But in this method, the pilot cell must be matched with *PV* array parameters which increases the system cost

(continued)

Table 1 (continued)

S. No.	MPPT technique	Remarks
10	Variable step size using commercial current mode control [21]	This method demonstrates an economically feasible method to control a current mode DC converter powered by *PV* panel. In the compensation pin of current mode DC regulator, a control voltage is injected to achieve MPPT. The advantage of this method is as they are small sized and easily available so the designing power system from *PV* system can be simplified
11	Fixed frequency finite-set model predictive control [22]	In this method, a model of the control system predicts the behavior of system and then an optimizer valuates the prediction results and finds out future control actions
12	Fuzzy logic control	The developed algorithm is able to track the maximum power with a convenient speed, and it shows a very dynamic response with sudden variations in environmental conditions. At the same time, the implementation of this algorithm is also possible with available components at a lower cost
13	Particle swarm optimization [23]	The swarm intelligence techniques are redundant, adaptable, extendable algorithm. In solar *PV* applications, PSO is used to adjust the duty cycle of DC converters to track the maximum power point. The swarm intelligence algorithm has the advantages to combine with other techniques and create a new approach for solving a problem
14	Neural network [24]	Neural networks find its best applications in the areas of nonlinear behavior or quite dynamic. It also provides analytical alternatives to conventional techniques
15	Genetic algorithm	GA can be combined with ANN and fuzzy logic techniques to expand the efficiency of the system performance A solution can be obtained by maximizing the fitness function. In genetic algorithm, the solution does not depend on initial conditions but works with a randomly generated population of individuals and chooses the best ones and hence it is able to search the exact maximum point
16	Artificial bee colony algorithm [25]	ABC algorithm is fairly recent member of swarm intelligence techniques. Based on natural behavior of honey bees in locating food resources, this algorithm develops a new methodology in search intelligence. Having the property of good accuracy and excellent tracking capability, this algorithm provides the facility of adjusting the duty cycle of the DC converter without using a linear controller

(continued)

Table 1 (continued)

S. No.	MPPT technique	Remarks
17	Cuckoo search algorithm [26]	This algorithm is one of the most recently developed. It is population-based algorithm, developed by simulating the intelligent breeding behavior of cuckoos through Levy's flight equation. This algorithm can be used as an optimization tool, or solving nonconvex and nonlinear problems
18	Firefly algorithm [27]	FA is based on flashing behavior of flies. This algorithm is based on assumption that since all the fireflies are unisex so they can be attracted to other fireflies irrespective of their sex and this attraction is proportional to the relative brightness; if there is no relative brightness, then each will move randomly. The landscape of the objective function finds the brightness of a firefly. For MPPT implementation, the brightness is simply proportional to the objective function

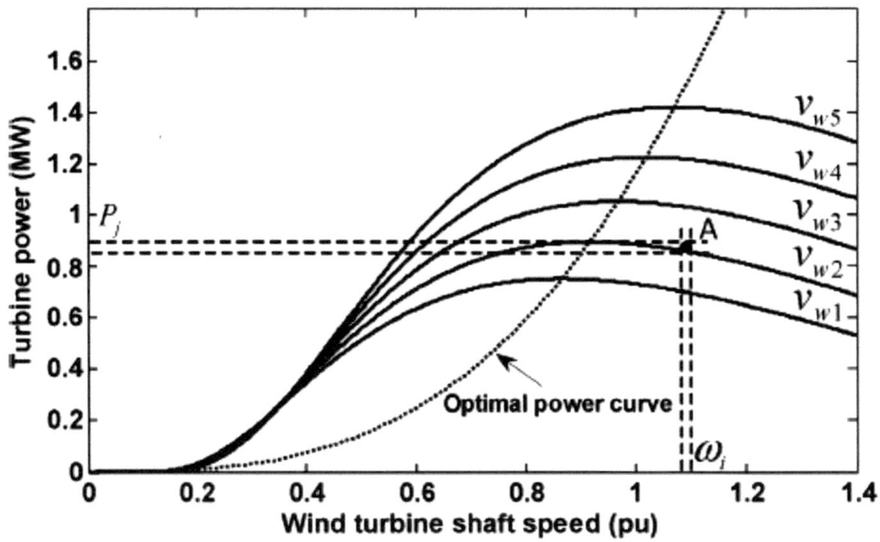

Fig. 3 MPPT characteristic of a wind turbine

4 Results

Sundareswaran [13] shows a comparison of variation of power, voltage, and current between P&O, PSO, and firefly algorithm for a *PV* system. From the graph, it can be clearly understood that the tracking speed of firefly algorithm is the fastest of all and has highest accuracy. Also, the steady state oscillations are fastest in firefly algorithm (Fig. 4).

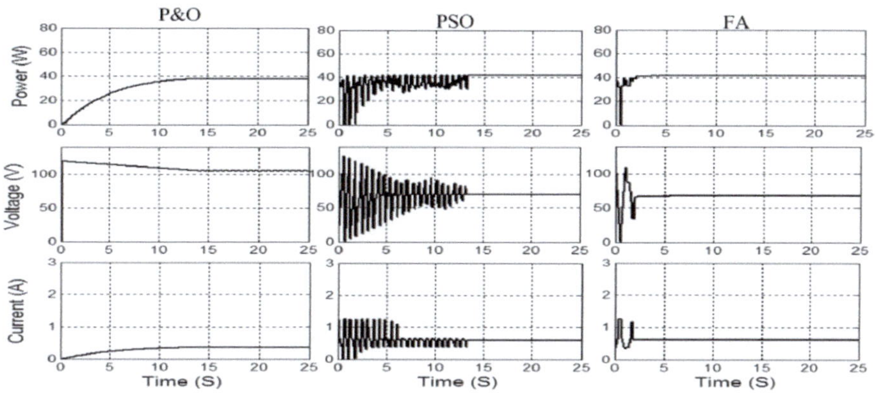

Fig. 4 Comparison of variation of power, voltage, and current between P&O, PSO, and firefly algorithm for a *PV* system [13]

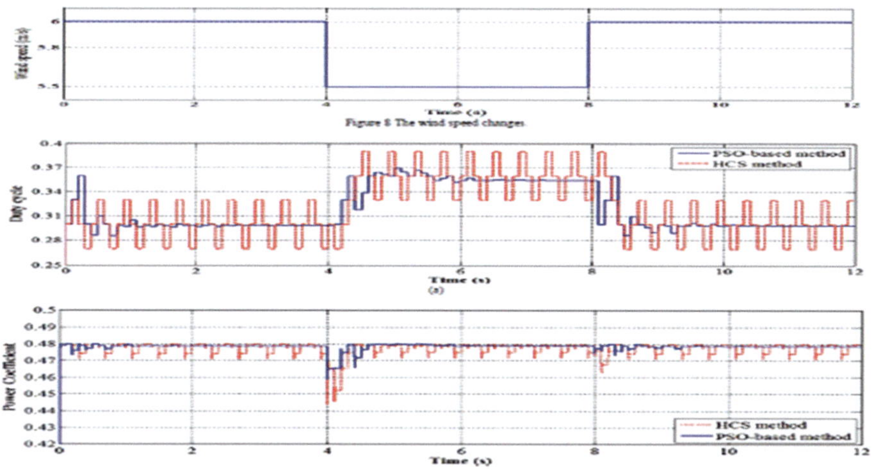

Fig. 5 PSO-based and HCS MPPT algorithms simulation results of **a** the duty cycle and **b** the power coefficient [14]

Abdullah [14] has shown implementation of swarm intelligence technique, i.e., particle swarm optimization to control the duty cycle of the boost converter. The converter provides the interface between wind turbine and the load by extracting maximum power from the wind. As shown from the result, the proposed PSO method has proven to be accurate and efficient to conventional hill climb search method (Fig. 5).

5 Conclusion

The classical optimization techniques for MPPT have numerous limitations on solving research models or mathematical models. The solution mechanisms of classical approaches are dependent on the type of function, i.e., linear or nonlinear and variable types, i.e., real or integer. The classical methods cannot be applied to problems involving different variable types and different objective functions. The conventional methods strictly follow sequential computations, produce precise answers, and require separate memory address for storing data. Hence, the above several limitations led to use of soft computing-based approaches. The soft computing strategies involve intelligent computational steps, and hence, computational time required is less.

Fuzzy systems and neural network systems relate the human way of thinking and interpretation. The genetic algorithms are related to the biological process in which the systems improve with time. The algorithms which are based on swarm intelligence techniques like ant colony optimization, particle swarm optimization, artificial bee colony optimization, cuckoo search algorithm imitate the intelligent behavior involved in animals, birds, insects, and microorganism. The particle swarm optimization is a special category of firefly algorithm. Many of the research papers have analyzed that firefly algorithm is superior in solving complex optimization problems. The firefly algorithm is more powerful in finding global optimum value in very less computing time. Due to some significant features, the cuckoo search algorithm provides faster convergence speed since it is based on Levy's flight equation.

The mentioned significance of the new optimization approaches can be used for maximum power point tracking systems in wind and solar energy systems. The cuckoo search algorithm and firefly algorithms can easily handle the shading conditions by their intelligent control approaches.

Apart from these research approaches, a combination of neuro-fuzzy, fuzzy genetic, and neurogenetic systems has also evolved to open new gateways for research. Each technique has their own merits and provides an efficient solution to the problems on different domains. This combination makes the system based on self-learning and decision making.

References

1. Nehrir, M.H., et al.: A review of hybrid renewable/alternative energy systems for electric power generation: configurations, control, and applications. IEEE Trans. Sustain. Energy **2**(4), 392–403 (2011)
2. Padma, S., Vijayalakshmi, K., Sangameshwaran, G.: Power generation using hybrid renewable energy resources for domestic applications. In: 2016 International Conference on Wireless Communications, Signal Processing and Networking (WiSPNET), Chennai, India, pp. 1993–1998 (2016)
3. Executive Summary Power Sector April 2016 (PDF) report. The Economic Times of India (2016). Retrieved 10 June 2016

4. Ministry of New and Renewable Energy, Annual Report 2015–2016
5. Scheme Wise Physical Progress in 2016–17. Ministry of New & Renewable Energy. 31 July 2016. Retrieved 31 Aug 2016
6. Muhamad, M. I., Radzi, M.A.M., Wahab, N.I.A., Hizam, H., Mahmood, M.F.: Optimal design of hybrid renewable energy system based on solar and biomass for halal products research institute, UPM. In: 2014 IEEE Innovative Smart Grid Technologies—Asia (ISGT ASIA), Kuala Lumpur, pp. 692–696 (2014)
7. Jainand, S., Agarwal, V.: A new algorithm for rapid tracking of approximate maximum power point in photovoltaic systems. IEEE Power Electron. Lett. **2**(1), 16–19 (2004)
8. Femia, N., Petrone, G., Spagnuolo, G., Vitelli, M.: Optimization of perturb and observe maximum power point tracking method. IEEE Trans. Power Electron. **20**(4), 963–973 (2005)
9. Linus, R.M., Damodharan, P.: Maximum power point tracking method using a modified perturb and observe algorithm for grid connected wind energy conversion systems. IET Renew. Power Gener. **9**(6), 682–689 (2015)
10. Kobayashi, K., Takano, I., Sawada, Y.: A study on a two stage maximum power point tracking control of a photovoltaic system under partially shaded irradiance conditions. In: IEEE Power Engineering Society General Meeting, pp. 2612–2617 (2003)
11. El-Khozondar, H.J., El-Khozondar, R.J., Matter, K., Suntio, T.: A review study of photovoltaic array maximum power tracking algorithms. https://doi.org/10.1186/s40807-016-00228
12. Wei, C., et al.: Reinforcement-learning-based intelligent maximum power point tracking control for wind energy conversion systems. IEEE Trans. Ind. Electron. **62**(10) (2015)
13. Sundareswaran, K., Peddapati, S., Palani, S.: MPPT of PV systems under partial shaded conditions through a colony of flashing fireflies. IEEE Trans. Energy Convers. **29**(2) (2014)
14. Abdullah, M.A., Yatim, A.H.M., Tan, C.W., Samosir, A.S.: Particle swarm optimization-based maximum power point tracking algorithm for wind energy conversion system. In: 2012 International Conference on Power and Energy, 2–5 Dec 2012
15. Noguchi, T., Togashi, S., Nakamoto, R.: Short-current pulse-based maximum-power-point tracking method for multiple photovoltaic-and-converter module system. IEEE Trans. Ind. Electron. **49**(1), 217–223 (2002)
16. Faranda, R., Leva, S.: Energy comparison of MPPT techniques for PV systems. WSEAS Trans. Power Syst. **3**(6) (2008)
17. Babaa, S.E., Armstrong, M., Pickert, V.: Overview of maximum power point tracking control methods for PV Systems. J. Power Energy Eng. **2**, 59–72 (2014)
18. Tse, K.K., Chung, H.S.H., Hui, S.Y.R., Ho, M.T.: A novel maximum power point tracking technique for PV panels. In: 2001 IEEE 32nd Annual Power Electronics Specialists Conference, PESC 2001, vol. 4, pp. 1970–1975 (2001)
19. Vicente, E.M., Moreno, R.L., Ribeiro, E.R.: MPPT technique based on current and temperature measurements. Int. J. Photoenergy **2015**, 9p (2015) (Article ID 242745) (Hindawi Publishing Corporation)
20. Kumar, C.K., Dinesh, T., Ganesh Babu, S.: Design and modelling of PV system and different MPPT algorithms. Int. J. Eng. Trends Technol. (IJETT) **4**(9) (2013)
21. Sheng, S., Lehman, B.: A simple variable step size method for maximum power point tracking using commercial current mode control DC-DC regulators. In: 2016 IEEE Applied Power Electronics Conference and Exposition (APEC), Long Beach, CA, pp. 2286–2291 (2016)
22. Abdel-Rahim, O., Funato, H., Haruna, J.: An efficient MPPT technique with fixed frequency finite-set model predictive control. In: 2015 IEEE Energy Conversion Congress and Exposition (ECCE), Montreal, QC, pp. 6444–6449 (2015)
23. Wang, L., Singh, C.: PSO-based multi-criteria optimum design of a grid-connected hybrid power system with multiple renewable sources of energy. In: Proceedings of the 2007 IEEE Swarm Intelligence Symposium (SIS 2007)
24. Hadji, S., Gaubert, J.P., Krim, F.: Experimental analysis of genetic algorithms based MPPT for PV systems. In: 2014 International Renewable and Sustainable Energy Conference (IRSEC), Ouarzazate, pp. 7–12 (2014)

25. Soufyane Benyoucef, A., Chouder, A., Kara, K., Silvestre, S., Aitsaheda, O.: Artificial bee colony based algorithm for maximum power point tracking (MPPT) for PV systems operating under partial shaded conditions. Appl. Soft Comput. **32**, 38–48 (2015) (1568-4946/© 2015 Elsevier)
26. Ahmed, J., Salam, Z.: A soft computing MPPT for PV system based on cuckoo search algorithm. In: 4th International Conference on Power Engineering, Energy and Electrical Drives Istanbul, Turkey, 13–17 May 2013
27. Niknam, T., Azizipanah-Abarghooee, R., Roosta, A.: Reserve constrained dynamic economic dispatch: a new fast self-adaptive modified firefly algorithm. IEEE Syst. J. **6**(4), 635–646 (2012)

Short-Term PV Power Forecasting Using Generalized Neural Network and Weather-Type Classification

Priyanka Chaudhary and M. Rizwan

Abstract Generation of electricity from solar energy is gaining huge attention because of advancement in the solar photovoltaic technology. Power from solar energy is intermittent in nature and requires a good forecasting method for efficient and reliable operation of smart grid systems. A large number of forecasting approaches are available in the literature. Due to intermittent nature of power obtained from sun, the results obtained from mathematical models for solar PV power prediction are not found satisfactory. An intelligent approach based on generalized neural network (GNN) is proposed and applied for the short-term solar PV power forecasting. Short-term forecasting for an hour to day ahead has applications in energy storage optimization, electricity pricing, etc. Keeping in mind aforesaid, 15 min ahead, short-term solar energy forecasting has been done and presented in this work for smart grid applications. The developed model requires an input of historical data set for PV output power, i.e. solar irradiance, temperature and the relative humidity of the site where solar PV is installed. The performance of the developed PV power forecasting model is evaluated with respect to the accuracy of the developed model for a 1 kWp practical system. Further, the evaluation of proposed method has been performed on the basis of root mean square error (RMSE) and mean absolute error (MAE).

Keywords Generalized neural network · Grid-integrated solar PV systems · Solar PV forecasting · Smart grid systems

1 Introduction

Solar photovoltaic (SPV) systems are playing important role in moving toward low-carbon energy resources. Output of solar photovoltaic depends on availability of solar

---

P. Chaudhary (✉) · M. Rizwan
Department of Electrical Engineering, Delhi Technological University, New Delhi, India
e-mail: priyankach.iilm@gmail.com

M. Rizwan
e-mail: rizwan@dce.ac.in

S. N. Singh et al. (eds.), *Advances in Energy and Power Systems*, Lecture Notes in Electrical Engineering 508, https://doi.org/10.1007/978-981-13-0662-4_2

irradiance and other meteorological parameters. Output power forecasting of SPV plant helps in controlling of variables in advance and optimize the capacity of energy storage system. A reliable and efficient forecasting model helps in improving grid-integrated operations, advance planning and maintenance. Prediction of solar energy becomes utmost important from the utilities point of view to balance the supply and demand for an efficient operation. Intermittent nature of power from solar energy becomes significant challenge for successful and economically efficient integration of solar power generating plants into utility grid. Hence, the prediction of SPV power output plays a significant role in sustainable power generation, power system operation and control. A large number of mathematical models for prediction of solar PV output power for cloudless sky conditions are available in the literature [1, 2]. Broad classification of short-term solar PV forecasting approaches is available in two categories, namely statistical methods and artificial intelligence methods. Some of statistical models are multiple linear regression, stochastic time series, autoregressive integrated moving average with exogenous variables (ARIMAX) and general exponential smoothing, state-space model and support vector regression (SVR). Intelligent models like artificial neural network, fuzzy inference systems, etc., are also available in the literature. The solar PV power output prediction using statistical models was not satisfactory subjected to the cloudy sky constraints. Further, fuzzy logic, neural network-based models are proposed by researchers to estimate the solar irradiance to deal with large uncertainties associated with the weather conditions [3]. In this paper, a generalized neural network (GNN)-based model for 15 min ahead solar PV output power forecasting has been developed and presented. Proposed model has the advantage of fast convergence and stronger training and learning ability. The developed model has been compared with the artificial neural network (ANN)-based model for results validation. Further, the evaluation of proposed method has been performed on the basis of root mean square error (RMSE) and mean absolute error (MAE).

2 Forecasting Model

2.1 Data Collection and Weather Classification

Data for global solar irradiance, ambient temperature, average relative humidity and past data for PV output power has been collected from Jamia Millia Islamia (JMI), National Institute of Solar Energy (NISE) and Ministry of New and Renewable Energy (MNRE), Government of India for New Delhi location at 15 min time interval and used as input for short-term solar PV output power forecasting. According to Indian weather scenario, seasonal classification of available solar power generation data is done as summer, winter and rainy seasons to achieve more granularity can be achieved to perform clustering on available historical data set. Input and output variables from the data set are not of the same order of magnitude. So, it is necessary to

Table 1 Neural network structure

Parameters	Value
Input parameter	4
Output	1
Input layer	1
Output layer	1
Hidden layer	1
Hidden layer neurons	3

normalize the data sets and convert it in the same order of magnitude. Normalization of the data within the range of (0.1–0.9) is used to convert actual monthly data, and this is done to improve convergence and learning process and given below:

$$L_s = \frac{Y_{max} - Y_{min}}{L_{max} - L_{min}}\left(L - L_{min}\right) + Y_{min} \tag{1}$$

where L is the actual solar irradiance data, L_s is the scaled solar irradiance data being used as input to the GNN model, L_{max} is maximum value of solar irradiance data in a particular column, L_{min} is minimum value of solar irradiance data in a particular column, Y_{max} is normalization upper range (0.9) and Y_{min} is normalization lower range (0.1).

2.2 ANN Model

The nonlinearity and complexity are the problems can be easily handled with the neural network [4, 5]. The neural network takes past data as feedback and generalizes the set of equations from previous calculations to the new ones. ANN does not require any physical link between solar power output and input parameters. Inputs to the model are global solar irradiance, ambient temperature, average relative humidity and previous PV output power. Parameters of structure of ANN are provided in Table 1. ANN model for short-term solar PV power output forecasting can be developed using the steps: input parameter selection; neural network architecture selection; training algorithm; training parameter selection. Backpropagation training algorithm for training of ANN has been used which helps in adjustment of the weight and threshold coefficients. NN training parameters are shown in Table 2. A single layer ANN structure has been used to develop the ANN-based solar PV output power forecasting model.

Table 2 Neural network training

Parameters	Value
Epochs	900
Momentum factor	0.1
Learning rate	0.003
Error tolerance	0.001

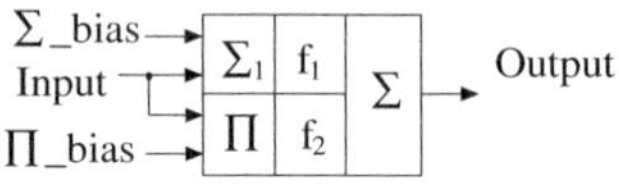

Fig. 1 Generalized neural network model

2.3 GNN Model

GNN is used to overcome problems associated with the performance, training and testing of ANN [6, 7]. It acts as a multi-layer feed-forward network in which each node performs a particular function on incoming signals as well as a set of parameters pertaining to this node. A common neuron structure has summation or product as the aggregation function with linear or nonlinear threshold function. A summation type GNN model is shown in Fig. 1, which shows the internal structure of GNN with summing up the outputs of the sigmoidal (f_1) and Gaussian (f_2) characteristic functions of product (Π) and summation (Σ_1) neuron [8].

The development of GNN model consisting of two phases: development phase and testing phase. The GNN model for solar PV power forecasting has been developed using same number of input and output variables as in ANN model. The output calculations can be divided into two parts: forward calculations and reverse calculations.

1. Forward calculations:

   (i) Output of the Σ_A part with a sigmoidal characteristic function can be calculated as

$$O_\Sigma = \frac{1}{1 + e^{-\lambda s * s\_net}} \tag{2}$$

   $s\_net = \Sigma W_{\Sigma i} X_i + X_{o\Sigma}$
   $\lambda s =$ gain scale factor of Σ_A.

   (ii) Output of the Π part with a sigmoidal characteristic function can be calculated as

$$O_\Pi = e^{-\lambda p * pi\_net^2} \tag{3}$$

   $pi\_net = \Pi W_{\Pi i} X_i * X_{o\Pi}$ and λp is the gain scale factor of Π_A.

   (iii) The output of the GNN is the function of two outputs and with the weights W and $(1 - W)$, respectively

$$O_{pk} = O_{\Pi} * (1 - W) + O_{\Sigma} W \tag{4}$$

2. Reverse calculations

(iv) After the calculation of output in forward mode, as in the feed-forward neural network, the output compared with the desired output to find the error. Back-propagation training algorithm is used here to train the GNN. Error of ith set of input for a single GNN is calculated by comparing the GNN output and desired output.

$$\text{Error } E_i = (Y_i - O_i) \tag{5}$$

(v) Sum squared error for the convergence of all the pattern is

$$E_p = 0.5 \sum Ei^2 \tag{6}$$

Multiplication factor of 0.5 has been taken to simplify the calculations.

vi) Weights associated with the Σ_A and Σ_B part of the summation type GN are:

$$W(k) = W(k-1) + \Delta W \tag{7}$$

$\Delta W = \eta \delta_k (O_{\Sigma} - O_{\Pi}) X_i + \alpha W(k-1)$
And $\delta_k = \sum (Y_i - O_i)$

(vii) Weights associated with the Σ_A part of the summation type GN are:

$$W \sum_i (k) = W \sum_i (k-1) + \sum W_{\Sigma i} \tag{8}$$

$\Delta W_{\Sigma i} = \eta \delta_{\Sigma i} X_i + \alpha W_{\Sigma i}(k-1)$
And $\delta_{\Sigma j} = \sum \delta_k W (1 - O_{\Sigma}) * O_{\Sigma}$
α momentum factor for better convergence and lies between 0 and 1
η learning rate and lies between 0 and 1

(viii) Weights associated with the Π part of the summation type GN are:

$$W\Pi_i(k) = W\Pi_i(k-1) + \Delta W_{\Pi i} \tag{9}$$

$\Delta W_{\Pi i} = \eta \delta_{\Pi i} X_i + \alpha W_{\Pi i}(k-1)$
And $\delta_{\Pi j} = \sum \delta_k (1 - W) * (-2 * pi\_net) * O_{\Pi}$

The output of summation part of generalized neural network can be obtained as:

$$O_{\Sigma} = f_1 \left(\sum W_{\Sigma i} X_i + X_{o\Sigma} \right) \tag{10}$$

The output of product part of generalized neural network can be obtained as:

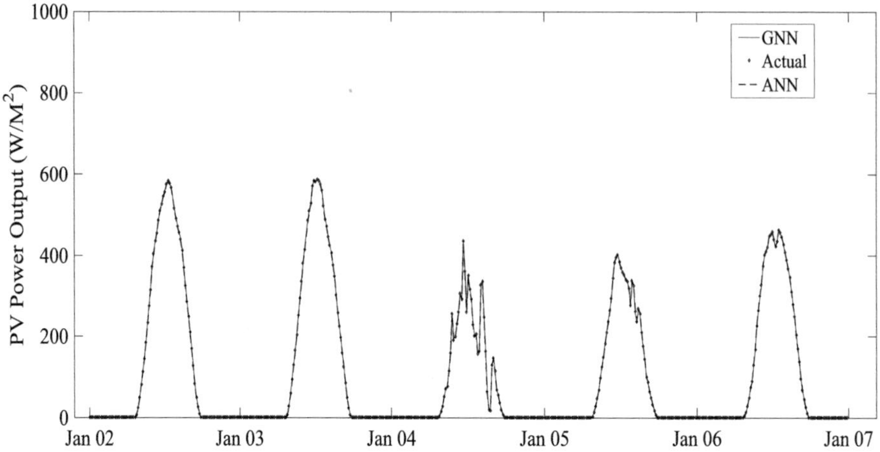

Fig. 2 Testing performance of ANN and GNN model for the winter season

$$O_{\Pi} = f_2(\Pi W_{\Pi i} X_i + X_{o\Pi}) \tag{11}$$

The final output of generalized neuron will be the sum of summation part and product part and can be mathematically written as:

$$O_i = O_{\Sigma} * W_{\Sigma} + O_{\Pi}(1 - W_{\Sigma}) \tag{12}$$

where O_{Σ} is output of the summation part of the neuron, O_{Π} is output of the product part of the neuron and W denotes weights.

3 Results and Discussions

The testing results for short-term solar power output forecasting using proposed GNN model are presented in this section and compared with ANN results. Data has been chosen and classified according to the three season types, namely summer, winter and rainy. Testing performances of developed models for winter season, summer season and rainy season are given in Figs. 2, 3 and 4, respectively. The average RMSE and percentage MAE for different seasons are given in Table 3. It is observed from Table 3 that the average values of RMSE and percentage MAE are less for GNN model as compared to ANN model. The average RMSE and percentage MAE are large for the rainy season because of the large uncertainties associated with the data.

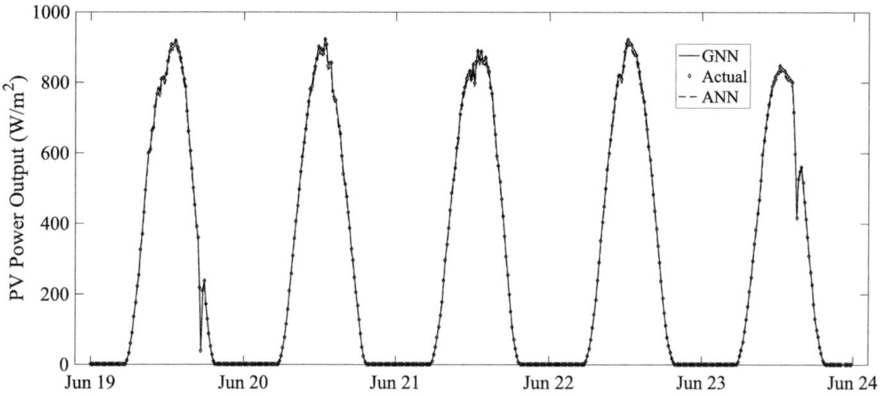

Fig. 3 Testing performance of ANN and GNN model for the summer season

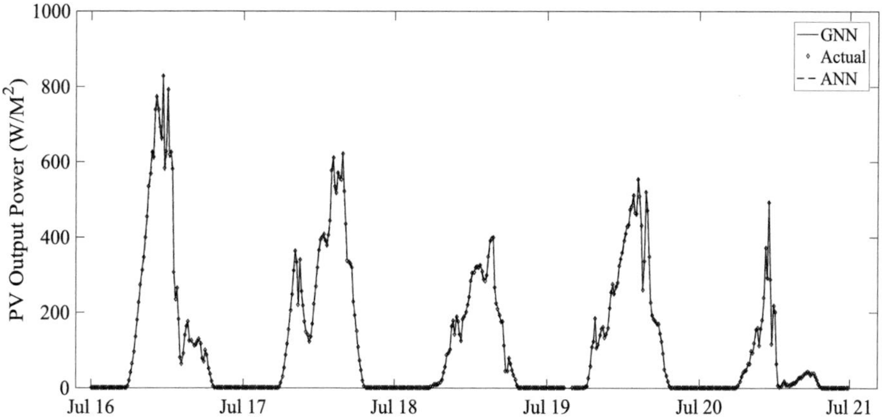

Fig. 4 Testing performance of ANN and GNN model for the rainy season

Table 3 Comparison of testing performances

Seasons	Model	RMSE(W)	MAE %
Summer	GNN	62.2	2.93
	ANN	72.8	3.34
Rainy	GNN	91.3	3.48
	ANN	104.2	4.42
Winter	GNN	69.2	3.05
	ANN	78.8	3.64

4 Conclusion

In this paper, 15 min ahead short-term solar PV output power forecasting models based on GNN has been developed and presented. The results of ANN and GNN are compared for solar power forecasting. The performance of the model is evaluated on the basis of statistical indicators such as MAE and RMSE. It is found that GNN performs better as compared to ANN in terms of learning time and convergence.

References

1. Rizwan, M., Jamil, M., Kothari, D.P.: Generalized neural network approach for global solar energy estimation in India. IEEE Trans. Sustain. Energy **3**(3), 576–584 (2012)
2. Rizwan, M., Jamil, M., Kothari, D.P.: Solar energy estimation using REST model for PV-ECS based distributed power generating system. Sol. Energy Mater. Sol. Cells **94**(8), 1324–1328 (2010)
3. Rizwan, M., Jamil, M., Kothari, D.P.: Generalized neural network methodology for short term solar power forecasting. In: 13th International Conference on Environment and Electrical Engineering (EEEIC), pp. 58–62. Wroclaw (2013)
4. Khotanzad, A., Afkhami-Rohani, R., Lu, T.L., Abaye, A., Davis, M., Maratukulam, D.J.: ANNSTLF-A neural-network-based electric load forecasting system. IEEE Trans. Neural Netw. **8**(4), 835–846 (1997)
5. Chaturvedi, D.K., Sinha, A.P., Chandiok, A.: Short-term load forecasting using soft computing techniques. Int. J. Commun. Netw. Syst. Sci. **3**, 273–279 (2010)
6. Ahmed, M.A., Ahmed, F., Akhter, M.W.: Estimation of global and diffuse solar radiation for Hyderabad, Pakistan. J. Basic Appl. Sci. **5**, 73–77 (2009)
7. Kalogirous, S.A.: Artificial neural networks in renewable energy systems applications: a review. Renew. Sustain. Energy Rev. **5**, 373–401 (2000)
8. Chaturvedi, D.K.: Soft computing techniques and its applications in electrical engineering. Springer, Berlin, Heidelberg, Germany (2008)

Model Order Reduction of Two Area Hydropower System Using Mixed Method

Sadaf Naqvi, Ibraheem Nasiruddin, Sana Ali and Shilpa Gupta

Abstract The complexity of power system models used for operation and control analysis has been increasing due to their operation in interconnected fashion. This paper presents a novel approach to develop reduced-order models of complex and large interconnected power system models by using mixed method of model order reduction technique. A state-space model of automatic generation control (AGC) system of a two-area interconnected power system model consisting of hydropower plants is developed. It is an original system of 11th order which is approximated to a reduced model of 3rd order. Validation of reduced-order model is done by comparing step response of original model with reduced-order model. The investigations of step responses show that the 3rd-order model is a good approximation of original model as far as its dynamic responses are concerned.

Keywords Mixed method · AGC · Model order reduction · Complex system Dynamic responses

1 Introduction

For the analysis of operational and control problems of present-day power systems, it is convenient to work with simpler mathematical models of reduced orders instead of using complex and higher-order models. Use of simple approaches and reduced models to carry out investigations are cost effective. A system can be simplified by developing lower-order models using various model order reduction techniques. Various methods of model order reduction have been proposed in the literature during last few decades. E.J. Davison has proposed a method for reducing large-order

S. Naqvi (✉) · S. Gupta
Inderprastha Engineering College, Ghaziabad, UP, India
e-mail: sadaf.naqvi@ipec.org.in

I. Nasiruddin
Qassim Engineering College, Qassim University, Buraydah, Saudi Arabia

S. Ali
IIIT, Delhi, India

S. N. Singh et al. (eds.), *Advances in Energy and Power Systems*, Lecture Notes in Electrical Engineering 508, https://doi.org/10.1007/978-981-13-0662-4_3

system models by constructing a matrix of lower order which has the same dominant eigenvalues and eigenvectors as that of the original system [1]. The procedure adopted by Davison suffers from two shortcomings as identified in [2]; first, it cannot be applied to singular matrix, and second, it does not exhibit the normally expected phenomenon due to ignoring the effect of unretained eigenvalues. Later, an improved version method was proposed by Davison [3].

Another novel technique based on aggregation was proposed by Aoki [4]. Using this technique, a reduced model of a system is derived by aggregating the original system state vector into a lower-dimensional vector. The aggregated model obtained from original model has been validated, and the output of aggregated model for a given input was found to be as near as that of original higher-order model. The aggregation matrix can be determined by modal aggregation and continued fraction approach [5]. Most of these model order reduction techniques use time domain-based techniques.

However, a variety of papers has been published based on frequency domain methods. Recently, moment matching technique is found to be very useful for this purpose [6]. In this method, a set of time functions of original higher-order model is matched with its simplified reduced-order model by selecting a number of relevant parameters. It has been demonstrated that there is no need to obtain time or frequency responses of original higher-order model [7] to achieve reduced-order models. The technique actually compares the time moments of the original higher-order model's impulse response with that of reduced-order model.

Another popularly known method based on frequency domain analysis is the Padé method of approximation. The Padé approximation approach is first used to reduce the SISO system, and later on, it is extended and used to reduce multivariable systems also [8]. First $(m+n)$ terms of power series expansion of original model are matched with the reduced-order model, where m is the order of numerator polynomial of original system and n is the order of denominator polynomial of original system.

An effective technique to reduce the complexity of a discontinuous system was presented by Bandyopadhyay et al. [9]. In this method, denominator coefficients of reduced model are obtained by the Routh table and numerator coefficients are obtained by the method of moment matching. This method suffers from a drawback of the requirement of complex computation. However, later, it is overcome in [10]. The method is further improved by Sastry et al. by introducing γ parameters. The denominator and numerator coefficients can be obtained using γ–δ parameters [11].

Other methods have also been suggested for the frequency-weighted model reduction problem [12, 13], but the approximation errors obtained using these techniques are large. To circumvent these, an improved technique for unweighted case is suggested in [14]. Recently, singular perturbation technique is applied to reduce complexity of system in [15]. In [16], a mixed method is proposed for deriving reduced-order models using the continued fraction approach and the Routh–Hurwitz array. This method is applicable to both single-input–single-output and multivariable systems and guarantees the stability of the reduced model when the original system is stable. This paper presents a reduced-order model using mixed method. However, the validity of this technique for large-order industrial systems is yet to be tested.

This paper considers a large and complex interconnected power system model for investigation. The dynamic model for its automatic generation control scheme is developed, and its reduced-order model is developed using mixed method. The step responses of the system model for both large-order original system and its reduced-order model are obtained and compared to demonstrate the validity of proposed technique for developing reduced-order models.

2 Description of Method

This method combines the feature of two model order reduction techniques: (i) continued fraction expansion approach as suggested by Chen and Shieh [17] and (ii) the Routh–Hurwitz array method as suggested by Krishnamurthy and Seshadri [18].

Let the nth-order system transfer function $G(s)$ and its rth-order reduced equivalent model transfer function $G_r(s)$ be represented as

$$G(s) = \sum_{j=1}^{n} a_{2,j} s^{j-1} / D(s) \tag{1}$$

and

$$G_r(s) = \sum_{j=1}^{r} b_{2,j} s^{j-1} / D_r(s) \tag{2}$$

where

$$D(s) = \sum_{j=1}^{n+1} a_{1,j} s^{j-1} \tag{3}$$

and $D_r(s)$ is the denominator polynomial of degree r.

Step 1 Form the Routh–Hurwitz stability array for the denominator polynomial as follows:

$$\begin{array}{llll}
a_{11} & a_{12} & a_{13}\ a_{14} & \cdots \\
a_{21} & a_{22} & a_{23}\ a_{24} & \cdots \\
a_{31} & a_{32} & a_{33}\ a_{34} & \cdots \\
\cdots & \cdots & \cdots\ \cdots & \cdots \\
a_{n-1,1} & a_{n-1,2} & & \\
a_{n,1} & & & \\
a_{n+1,1} & & &
\end{array}$$

A polynomial of lower order r may easily be constructed with the help of above array, to give

$$D_r(s) = a_{(n+1-r),1}s^r + a_{(n+2-r),1}s^{r-1} + a_{(n+1-r),2}s^{r-2} + \ldots \quad (4)$$

Now compare Eq. (4) with the following expression to get $b_{1,j}$ coefficients

$$\sum_{j=1}^{r+1} b_{1,j}s^{j-1} \text{ with } b_{1,r+1} = 1 \quad (5)$$

Step 2 Evaluate the quotients h_i, $i = 1, 2, \ldots, r$ by using the Routh algorithm

$$\left.\begin{aligned} a_{i,j} &= a_{i-2,j+1} - h_{i-2}a_{i-1,j+1} \\ i &= 3, 4, \ldots; j = 1, 2, \ldots \\ h_i &= \tfrac{a_{11}}{a_{i+1,1}}; \end{aligned}\right\} \quad (6)$$
$$i = 1, 2, \ldots, r \text{ provided } a_{i+1,1} \neq 0$$

Step 3 Determine b_{2j} of (2), by applying the following reverse algorithm:

$$b_{i+1,j+1} = b_{i,1}/h_i;$$

$$\left.\begin{aligned} b_{i+1,1} &= \tfrac{b_{i,1}}{h_i};\ i = 1, 2, \ldots r \\ b_{i+1,j+1} &= \tfrac{(b_{i,j+1} - b_{i+2,j})}{h_i}; \\ i = 1, 2, \ldots r - j;&\quad j = 1, 2, \ldots r - 1 \end{aligned}\right\} \quad (7)$$

These steps are repeated for each numerator.

3 Application on AGC Two-Area Model

Power system model considered for the application purpose is a two-area interconnected power system consisting of identical power plants with hydroturbines [19].

The state-space equation of power system is written as:

$$\mathrm{d}x/\mathrm{d}t = Ax + Bu$$
$$y = Cx + Du$$

The system matrices A, B, C, and D described by the equations for a closed-loop system can be written as:

$$A = \begin{bmatrix} 0.15 & 6.97 & 2.01 & 0.05 & 0.59 & 0.68 & 0.46 & 0.01 & -7.09 & 0.36 & 0.18 \\ 0.17 & -2.0 & 2.04 & 0.36 & 0 & 0 & 0 & 0 & 0 & 0 & 0 \\ -0.14 & -0.55 & -2.23 & -0.12 & -0.06 & 0.31 & 0.33 & 0.05 & 1.22 & -0.36 & 0.27 \\ -0.81 & 0 & 0 & -1.95 & -3.06 & 2.37 & -2.51 & -0.07 & 11.02 & -1.66 & 1.05 \\ 0 & 0 & 0 & 0 & 0.17 & -2.00 & 2.04 & 0.36 & 0 & 0 & 0 \\ -0.21 & 0.28 & 5.85 & -0.39 & -1.24 & -3.14 & -2.75 & -0.46 & -3.28 & 0.87 & -1.74 \\ 0 & 0 & 0 & 0 & -0.81 & 0 & 0 & -1.95 & 0 & 0 & 0 \\ 0.55 & 0 & 0 & 0 & -0.55 & 0 & 0 & 0 & 0 & 0 & 0 \\ 0.43 & 0 & 0 & 0 & 0 & 0 & 0 & 0 & 1 & 0 & 0 \\ 0.13 & 0 & 0 & 0 & 0 & 0 & 0 & 0 & 1 & 0 & 0 \\ 0 & 0 & 0 & 0 & 0.43 & 0 & 0 & 0 & -1 & 0 & 0 \end{bmatrix};$$

$$B = \begin{bmatrix} -0.4003 & 0.0 \\ 0.0 & 0.0 \\ 2.001 & -0.4003 \\ 0.0 & 0.0 \\ 1.949 & 0.2001 \\ 0.0 & 0.0 \\ 0.0 & 1.949 \\ 0.0 & 0.0 \\ 0.0 & 0.0 \\ 0.0 & 0.0 \\ 0.0 & 0.0 \end{bmatrix};$$

$$C = \begin{bmatrix} 1\,0\,0\,0\,0\,0\,0\,0\,0\,0\,0 \\ 0\,0\,0\,0\,1\,0\,0\,0\,0\,0\,0 \\ 0\,0\,0\,0\,0\,0\,0\,0\,1\,0\,0 \end{bmatrix}; \quad D = [0]$$

Its transfer function can be represented as:

$$G_I = \frac{\begin{bmatrix} g_{11}(s) & g_{12}(s) \\ g_{21}(s) & g_{22}(s) \\ g_{31}(s) & g_{32}(s) \end{bmatrix}}{D_n(s)}$$

where

$$\begin{aligned} g_{11} &= 0.21s^9 - 0.5644s^8 - 13.46s^7 - 27.28s^6 + 60.25s^5 + 276.7s^4 \\ &\quad + 383.5s^3 + 238.1s^2 + 54.17s - 6.57e - 15 \\ g_{12} &= -0.4003s^{10} - 4.82s^9 - 17.26s^8 + 17.49s^7 + 322.6s^6 + 1090s^5 \\ &\quad + 1848s^4 + 1699s^3 + 795.1s^2 + 161.8s - 9.4e - 16 \\ g_{21} &= 0.2001s^{10} + 1.336s^9 + 18.11s^8 + 118.1s^7 + 378.8s^6 + 788.3s^5 \\ &\quad + 1137s^4 + 952.9s^3 + 364.3s^2 + 54.17s - 8.38e - 15 \end{aligned}$$

$$g_{22} = 1.949s^{10} + 23.37s^{9} + 125s^{8} + 388.2s^{7} + 811.4s^{6} + 1338s^{5} + 1760s^{4} + 1541s^{3} + 752.8s^{2} + 161.8s + 2.512e - 014$$

$$g_{31} = -0.1091s^{9} - 0.616s^{8} - 10.18s^{7} - 71.7s^{6} - 221.3s^{5} - 396.8s^{4} - 469.1s^{3} - 310.3s^{2} - 68.76s - 7.428e - 016$$

$$g_{32} = -1.28s^{8} - 15.36s^{7} - 77.55s^{6} - 202s^{5} - 266.4s^{4} - 135s^{3} + 48.42s^{2} + 86.03s + 23.02$$

$$D_n = s^{11} + 15.79s^{10} + 117.4s^{9} + 547.7s^{8} + 1765s^{7} + 4053s^{6} + 6662s^{5} + 7718s^{4} + 6045s^{3} + 2995s^{2} + 846.4s + 102.5$$

Step 1 Form the Routh array from denominator polynomial
To find the third-order reduced model, denominator polynomial can be written from the Routh array starting from s^3 row:

$$2004.1s^3 + 1298.1s^2 + 510.8s + 102.5$$

$$D_r(s) = s^3 + 0.64s^2 + 0.25s + 0.005$$

This gives $b_{11}=0.005$; $b_{12}=0.25$; $b_{13}=0.64$; $b_{14}=1$.

Step 2 Find the quotients h_i by using the Routh algorithm for each numerator. First row of the following array is formed from the denominator, and second row is from first numerator, i.e., from g_{11}
Let the coefficients of array be represented by $a_{i,j}$, and the values of coefficients and h_i are determined by Eq. (6).

Step 3 Find the values of reduced numerator using relation (7)

$$b_{21} = b_{11}/h_1; \quad b_{31} = b_{21}/h_2;$$

$$b_{41} = b_{31}/h_3 \;\&\; b_{22} = (b_{12} - b_{31})/h_1; \quad b_{23} = (b_{13} - b_{32})/h_1$$

Reduced numerator will be given by

$$g_{11r} = b_{21} + b_{22}s + b_{23}s^2$$

$$g_{11r} = -3.28e - 18 + 0.027s - 4.15e - 17s^2$$

Repeat the Steps 2 and 3 for g_{12}, g_{13}, g_{21}, g_{22} and g_{23}

$$g_{12r} = 0.48s^2 + 0.08s - 4.7e - 18$$

$$g_{21r} = 0.09s^2 + 0.027s - 4.18e - 018$$

$$g_{22r} = 1.58e - 016s^2 + 0.08s + 1.25e - 017$$

$$g_{31r} = -4.69e - 18s^2 - 0.03s - 3.71e - 19$$

$$g_{32r} = -0.004s^2 + 5.32e - 03s + 0.0115$$

Transfer function of reduced third-order model is found to be

Table 1 Set of closed-loop system eigenvalues

Eigenvalues of original model	Eigenvalues of third-order reduced model
−1.2280+2.5972*j*	
−1.2280 − 2.5972*j*	
−3.4901	−0.15369 + 0.35588*j*
−1.1019 + 1.4569*j*	−0.15369 −0 .35588*j*
−1.1019 −1 .4569*j*	−0.34036
−2.8600	
−2.0709	
−1.3134	
−0.50121 + 0.31513*j*	
−0.50121 − 0.31513*j*	
−0.39129	

$$G_r(s) = \frac{\begin{bmatrix} g_{11r} & g_{12r} \\ g_{21r} & g_{22r} \\ g_{31r} & g_{32r}) \end{bmatrix}}{D_r(s)}$$

To test the system stability margins available for the models, a set of closed-loop system eigenvalues are obtained which are placed in Table 1.

4 Results

A reduced-order model is obtained as a complex 11th-order AGC system of an interconnected power system. For the validation of reduced-order model, dynamic step responses of original and reduced-order models are obtained and compared. These response plots are shown in Figs. 1, 2 and 3. Few important parameters as frequency deviation and change in tie line power are plotted for both original and approximated models. Figures 1 and 2 show the response of frequency deviation of area 1 and area 2, respectively. A step disturbance is applied at one area, and variation of frequency deviation is observed. It is observed from the responses that peak overshoot of original model is higher than that of reduced model; however, the steady-state response of both the systems is comparable. The investigations of response plots of Figs. 1 and 2 reveal that the oscillations have increased in reduced-order system. Figure 3 shows that the reduced model follows the same pattern as that of original model. However, oscillations are increased. From the inspection of closed-loop system eigenvalues, it is revealed that both the original and reduced-order models are stable.

Fig. 1 Frequency deviation of area 1

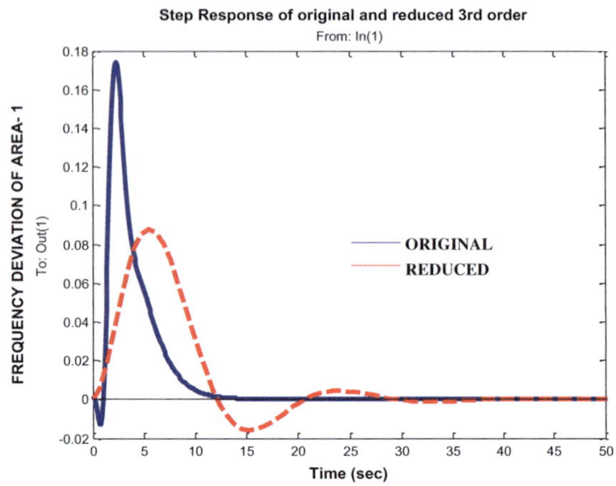

Fig. 2 Frequency deviation of area 2

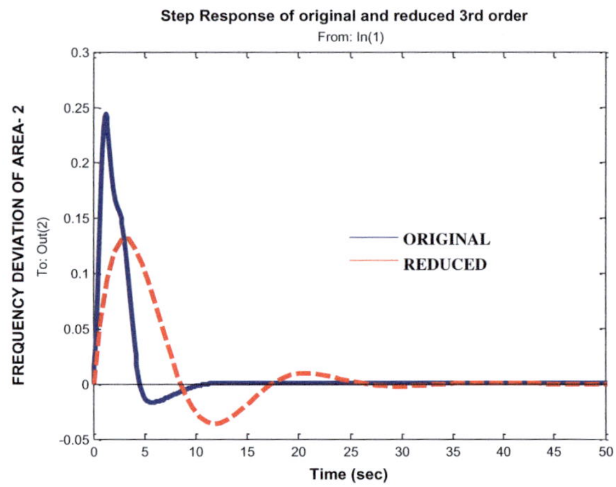

Fig. 3 Change in tie line power

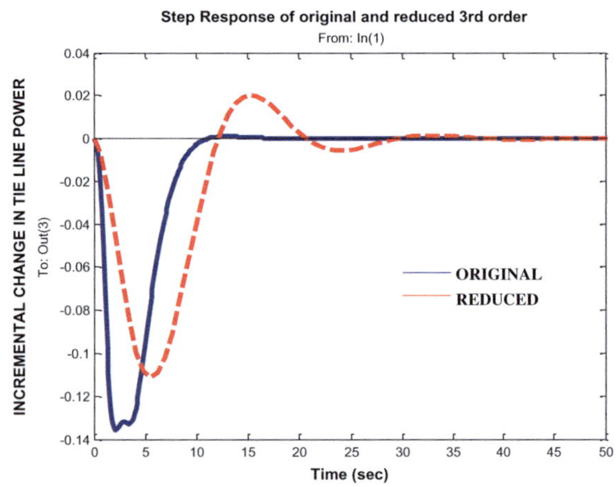

5 Conclusions

This paper shows that the complex system can be converted to simplified system using mixed method. Simplified system is a reduced-order model of the original system. Reduced-order model is validated by comparing its step response with the original one. The stability of the proposed model is investigated by obtaining the eigenvalues. From the results obtained, it is concluded that proposed approximated/simplified model gives a better understanding of the system; moreover, the hardware requirement to synthesize a reduced model will be appreciably less.

References

1. Davison, E.J.: A method of simplifying linear dynamic systems. IEEE Trans. Automat. Contr. **AC-11**, 93–101 (1966)
2. Chidambara, M.R.: On a method for simplifying linear dynamic systems. IEEE Trans. Automat. Contr. 119–120 (1967)
3. Davison, E.J.: A new method for simplifying large linear dynamic systems. IEEE Trans. Auto. Contr. 214–215 (1968)
4. Aoki, M.: Control of large-scale dynamic systems by aggregation. IEEE Trans. Auto. Contr. **AC 13**(3), 246–253 (1968)
5. Lamba, S.S., Vittal Rao, S.: Aggregation matrix for the reduced order continued fraction expansion model of Chen & Shieh. IEEE Trans. Auto. Contr. **23**(1), 81–83 (1978)
6. Ionescu, T.C., Astolfi, A.: Nonlinear moment matching-base model order reduction. IEEE Trans. Autom. Contr. **61**(10), 2837–2847 (2016)
7. Bosley, M.J., Lees, F.P.: Methods for the reduction of high order state variable models to simple transfer functions. In: IFAC Symposium on Digital Simulation of Continuous Processes, Gyor (1971)
8. Shamash, Y.: Linear system reduction using Pade approximation to allow retention of dominant modes. Int. J. Contr. **21**(2), 257–272 (1975)
9. Bandyopadhyay, B., Ismail, O., Gorez, R.: Routh Pade approximation for interval systems. IEEE Trans. Auto. Contr. **39**(12), 2454–2456 (1994)
10. Bandyopadhyay, B., Upadhye, A., Ismail, O.: γ–δ Routh approximation for interval systems. IEEE Trans. Auto. Contr. **42**(8), 1127–1130 (1997)
11. Sastry, G.V.K.R., Raja Rao, G., Mallikarjuna Rao, P.: Large scale interval system modeling using Routh approximants. Electron. Lett. **36**(8), 768–769 (2000)
12. Enns, D.: Model reduction with balanced realizations: an error bound and a frequency weighted generalization. In: Proceedings of 23rd IEEE Conference Decision Control, Las Vegas, NV, pp. 127–132 (1984)
13. Wang, G., Sreeram, V., Liu, W.: A new frequency-weighted balanced truncation method and an error bound. IEEE Trans. Automat. Control **44**(9), 1734–1737 (1999)
14. Moore, B.C.: Principal component analysis in linear systems: controllability, observability and model reduction. IEEE Trans. Auto. Contr. **AC-26**(1), 17–32 (1981)
15. Tang, Y., Prieur, C., Girard, A.: Singular perturbation approximation of linear hyperbolic systems of balance laws. IEEE Trans. Auto. Contr. **61**(10), 3031–3037 (2016)
16. Jayant, P.: System reduction by a mixed method. IEEE Trans. Auto. Contr. **AC 25**(5), 973–976 (1980)
17. Chen, C.F., Shieh, L.S.: A novel approach to linear model simplification. Int. J. Contr. **8**, 561–570 (1968)

18. Krishnamurthi, V., Seshadri, V.: A simple and direct method for reducing order of linear systems using Routh approximations in the frequency domain. IEEE Trans. Auto. Contr., 797–799 (1976)
19. Kumar, P., Ibraheem.: Dynamic performance evaluation of 2-area interconnected power systems: a comparative study. J. Instit. Eng. (India) **78**, 199–209 (1998)

Current Trends in Control Techniques in Renewable Energy: A Review

Srishti, Prerna Gaur and Surabhi Chandra

Abstract This manuscript is a review work of various maximum power point tracking (MPPT) techniques which experience partial shading conditions (PSCs), an unavoidable complication that significantly reduces the efficiency of the overall system. The exhaustive comparison of various MPPT's taken by the researchers which tracks the global peak (GP) of a photovoltaic (PV) array under PSC is compared and the best available option is proposed for researchers. MPPT techniques such as perturb and observe (P&O), improved particle swarm optimization (IPSO), and grey wolf optimization (GWO) are the recent techniques proposed by the researchers. Various DC–DC converter topologies with PV system are also compared for the purpose of maximum power point tracking.

Keywords Maximum power point tracking (MPPT) · Grey wolf optimization (GWO) · Partial shading conditions (PSC)

List of Abbreviations

(1) Maximum power point tracking (MPPT)
(2) Partial shading conditions (PSC)
(3) Global peak (GP)
(4) Photovoltaic (PV)
(5) Perturb and observe (P&O)
(6) Improved particle swarm optimization (IPSO)
(7) Grey wolf optimization (GWO)

Srishti (✉) · P. Gaur · S. Chandra
Netaji Subhas Institute of Technology, Delhi, India
e-mail: srishti_26@yahoo.com

P. Gaur
e-mail: prernagaur@yahoo.com

S. Chandra
e-mail: surabhichandra.en@gmail.com

S. N. Singh et al. (eds.), *Advances in Energy and Power Systems*, Lecture Notes in Electrical Engineering 508, https://doi.org/10.1007/978-981-13-0662-4_4

(8) Artificial intelligence (AI)
(9) Particle swarm optimization (PSO)
(10) Insulated-gate biploar transistor (IGBT)
(11) Single-ended primary-inductor converter (SEPIC Converter)

1 Introduction

The sunlight is directly converted into electricity by photovoltaic system. The energy that obtained depends on various factors such as solar radiation, temperature, and the voltage produced [1]. There is a mismatch between the module currents of Photovoltaic (PV) array because PV array modules receive non-uniform solar irradiation in case of partial shading conditions (PSCs). The main reason of the degradation of power generation of a PV array is partial shading [2]. Various conventional maximum power point tracking (MPPT) methods are reported in the past. These MPPT methods show best results only under constant radiances. In this review paper, the focus of the authors is to highlight the comparison in conventional and recent artificial intelligence (AI)-based MPPTs in order to improve the issues of fluctuations and efficiency generated in the output of PV system by tracking global peak (GP) during partial shading conditions (PSCs) [2]. MPPT techniques are required for extracting the maximum available power from PV systems. MPPT algorithms are applied to generate duty cycle in power electronic converters connected to PV systems to track the voltage and current for maximum power generation [3]. The analysis of various DC–DC converter topologies is also emphasized in this paper.

2 PV System Characteristics During PSC

2.1 *Characteristic Feature of PV Cell*

Equivalent single diode model can be used to represent a PV cell. Following symbolic notations have been used in the PV model.

I_{pv} Current source of PV module;
D Diode allied in parallel to current source;
R_s Sum of resistances due to all the components that comes in path of current;
R_p Represent the leakage across the P–N junction which is required to be as high as possible;
I Difference between the photocurrent I_{pv} and the diode current, which is specified by:

$$I = I_{pv} - I_0\left[\exp\left(\frac{qV + qR_s}{N_s K_s T_a} - 1\right)\right] - \frac{V + R_s I}{R_p} \tag{1}$$

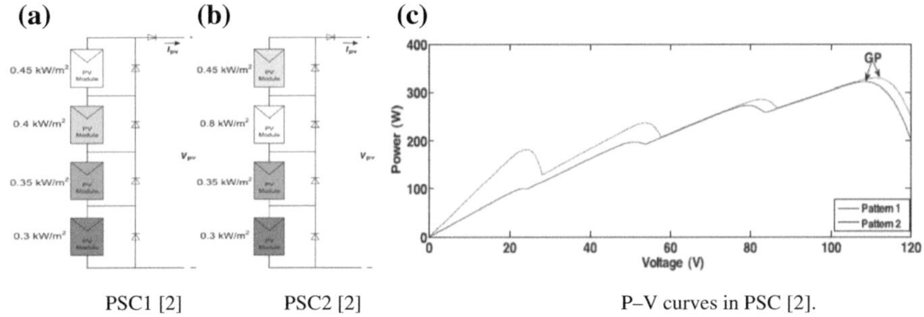

Fig. 1 4S configuration during various PSC. **a** PSC1 [2], **b** PSC2 [2], **c** P–V curves in PSC [2]

where I_0 is shown as saturation current, K_s is the Boltzmann constant, q is given as electron charge, T_a is temperature in kelvin, and N_s is the cells connected in series [2].

2.2 *PV and PSC Elementary Description*

Sunlight is the most desirable source of renewable clean and abundant energy. PV cells are used to convert sunlight into electricity. Numerous PV modules are connected in series or parallel in a PV cell as shown in Fig. 1a, b. PV cells give different output during PSC. Bypass diodes are present in PV due to which characteristic curve of PV yields a multiple peak curve during PSC condition as shown in Fig. 1c. In this review paper, the results of two PV configurations are presented by the researchers comprising of four modules allied in series (4S configuration) having shading patterns of two different types [2]. As mentioned the P–V curves of two shading patterns are as shown in Fig. 1.

3 MPPT Algorithms

3.1 *Perturb and Observe (P&O) Algorithm*

P&O system is an example of hill climbing method. Firstly, a slight perturbation is acquainted with the P&O algorithm. The introduction of this perturbation leads to the change in the power of the module. As soon as the maximum power is obtained, the power at the next interval of time decreases and henceforth the perturbation moves in opposite direction. When the maximum point is reached, the algorithm will oscillate around that point. The researches [1] have shown that perturbation size should be kept small so as to keep the variation of power small as shown in Fig. 2.

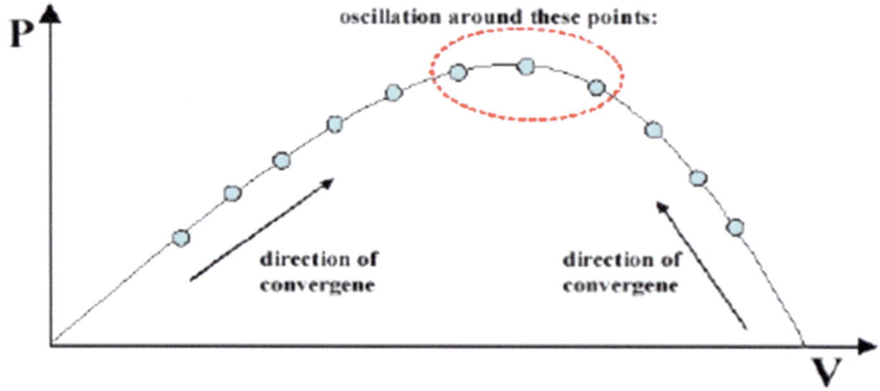

Fig. 2 Perturb and observe algorithm

3.2 *Improved Particle Swarm Optimization (IPSO) Algorithm*

In IPSO algorithm, the optimal and sub-optimal locations, which each particle encountered and the swarm meted, are kept. Based on these locations, the IPSO produces four velocities for each particle and obtains the particle's iterative position. The IPSO enlarges the search space and enhances global search ability. This review work also demonstrates IPSO algorithm which has been anticipated in order to overcome the limitation of particle swarm optimization (PSO). The main disadvantage of PSO algorithm is that it converges in local optima. Thus, the algorithm known as IPSO was proposed with more proficient global searching ability. In order to overcome this local optima problem, the improved algorithm has adopted a new mechanism. Initially, the exact location of the finest solution is not known by the particle when it is searching in the solution space. But IPSO helps to know the best locations and also it helps to record the worst locations of the individual particle. Hence, this can help in making individual particles aware of the worst locations which can help in widening the global searching space of particles. The particle velocity and position renewal formula are given in (2) and (3):

$$V_{id} = \omega V_{id}\eta_1 \text{rand}()(X_{id} - P_{idw}) + \eta_2 \text{rand}()\left(X_{id} - P_{gdw}\right) \tag{2}$$

$$X_{id} = X_{id} + V_{id} \tag{3}$$

where V_{id} is the particle velocity, ω is called inertia weight, η_1 and η_2 are the constants and are known as accelerating factors, rand() are random numbers, P_{idw} and P_{gdw} are given as the best position particle id has found and the best particles position in the entire swarm [4].

3.3 Grey Wolf Optimization (GWO) Algorithm

The GWO algorithm imitated the hunting mechanism and leadership pyramid of grey wolves. Grey wolves prefer to live in a group. Four types of grey wolves are employed for simulating the leadership pyramid. The types of grey wolves are alpha (α), beta (β), delta (δ), and omega (ω). While designing GWO, alpha (α) is considered as the fittest of all solution so as to mathematically model the social pyramid. Consequently, the next two best solutions are named as beta (β) and delta (δ), respectively, and remaining solutions are considered as omega (ω). Three stages of GWO algorithm are shown in Fig. 3, viz. hunting and tracing for prey, surrounding prey, and attacking prey are the steps to perform GWO algorithm. The encircling behaviour of Grey wolves is shown in (4) and (5).

$$\vec{D} = \left|\vec{C} \cdot \overrightarrow{X_p}(t) - \overrightarrow{X_p}(t)\right| \tag{4}$$

$$\vec{X}(t+1) = \overrightarrow{X_p} - \vec{A} \cdot \vec{D} \tag{5}$$

where t is current iteration, D, A, and C denote vector coefficient, X_p is given as prey position vector, and X shows the grey wolf position vector. Equations (6) and (7) show the vectors A and C:

$$\vec{A} = 2\vec{a} \cdot \overrightarrow{r_1} - \vec{a} \tag{6}$$

$$\vec{C} = 2 \cdot \overrightarrow{r_2} \tag{7}$$

Fig. 3 Grey wolves' behaviour of hunting. **a** Tracing prey, **b–d** surrounding prey, **e** attacking prey

Table 1 Comparison of the performance of various MPPT methods for 4S configuration

Shading pattern	Maximum power from PV curve (W)	Tracking techniques	Maximum power (W)	Maximum voltage (V)	Maximum current (A)	% Tracking efficiency
1	320	P&O	100.2	24.2	4.14	31.30
		IPSO	319.2	110.52	2.888	99.75
		GWO	319.4	110.55	2.889	99.81
2	330	P&O	180	23.07	7.80	54.54
		IPSO	329.5	112.3	2.934	99.84
		GWO	329.6	112.3	2.934	99.87

During iterations the components linearly decrease from 2 to 0 and are given as random vectors in [0, 1]. Alphas are known as the leaders which usually guide the hunting. But sometimes beta and delta wolves may also participate in hunting. Care of the wounded wolves is guided by Delta and omega. Therefore, as a candidate solution alpha is preferred [2].

4 Comparison of P&O, IPSO, and GWO Algorithm

The review of above all algorithms shown was implemented under PSC and also on the varying insolation level for 4S configuration. In order to design single diodes model of a PV module the following parameters are taken:

$P_{\max} = 200$ W, $V_{OC} = 32.8$ V, $I_{SC} = 8.21$ A, $V_{mp} = 26.3$ V, and $I_{mp} = 7.61$ A. The parameters of IPSO are $w_{\max} = 1$, $w_{\min} = 0.1$, $c_{1,\max} = 2$, $c_{1,\min} = 1$, $c_{2,\max} = 2$, $c_{1,\min} = 1$. Figure 4, [2] shows the comparison of various 4S configuration graphs of power, voltage, and current of GWO-, IPSO-, and P&O-based MPPT (Table 1).

5 Various DC–DC Converters

The review work also highlights various converters as below.

5.1 CUK Converter

CUK converter is a DC–DC converter in which the output voltage can be greater or lesser than the given input voltage, but the output voltage has a reverse polarity of

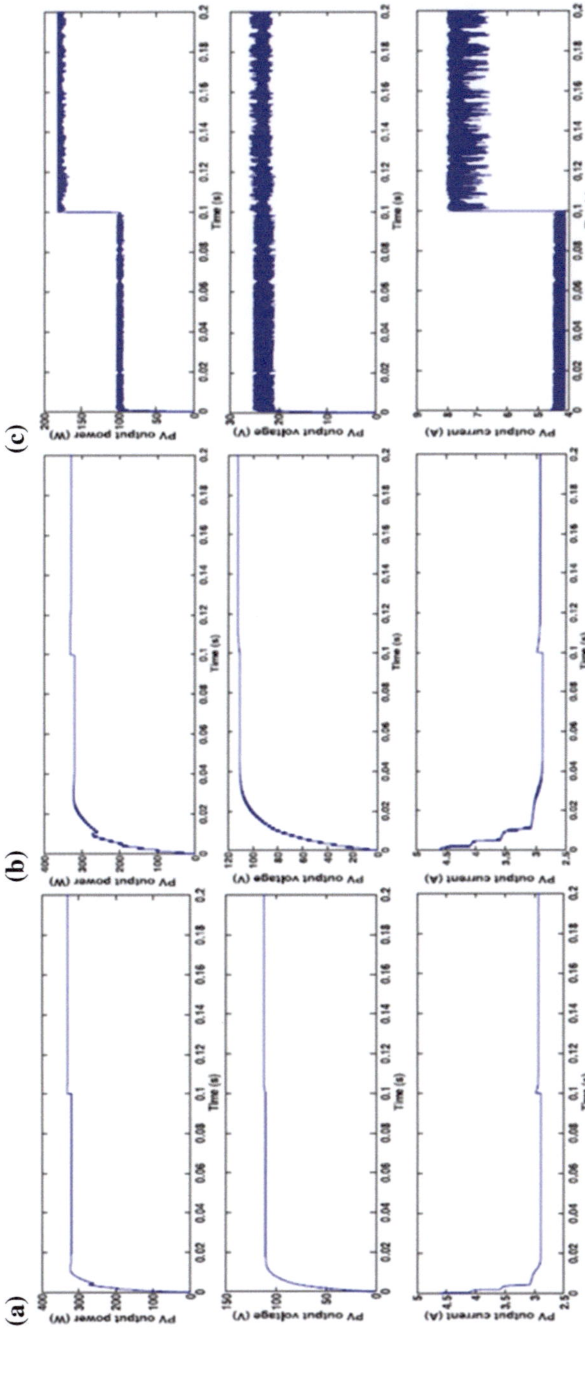

Fig. 4 Graphs for 4S configuration. **a** GWO MPPT, **b** IPSO MPPT, **c** P&O MPPT

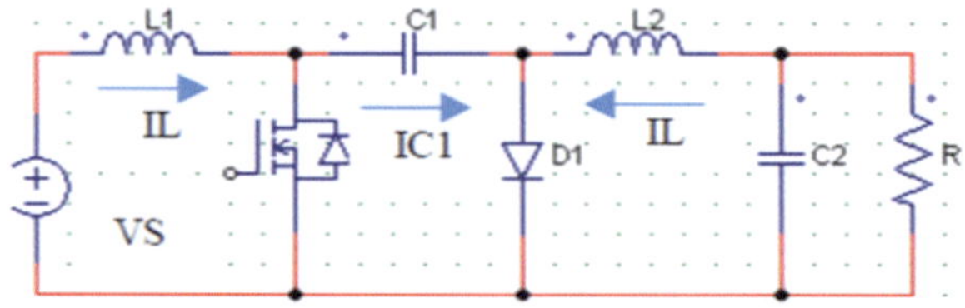

Fig. 5 CUK converter switching topology

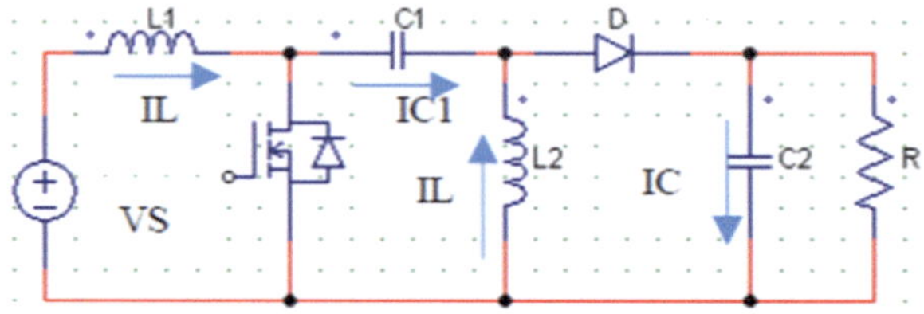

Fig. 6 SEPIC converter switching topology

the input voltage. CUK switching topology is shown in Fig. 5. Inductor component L_1 functions as a filter in order to prevent larger harmonics. Capacitor C_1 helps in transferring energy to inductor of the CUK converter. The voltage across inductor is zero at steady state operation $C_1 = V_s - V_0$. When the switch is closed the condition for diode is forward biased and current on C_1 is given by $\left(IC_{1(\text{close})} = -IL_1\right)$. Also the current in L_1 and L_2 flows through the diode when the switch is open. Current on C_1 is given as $\left(IC_{1(\text{open})} = -IL_1\right)$. Power absorbed in load R is equal to the power delivered from the source $-V_0IL_2 = V_sIL_1$ [5–7].

5.2 *Single-Ended Primary-Inductor Converter (SEPIC Converter)*

The SEPIC is a type of DC/DC converter allowing voltage at its output to be greater than, less than, or equal to that at its input. There is quite similarity between SEPIC and CUK converter. The only dissimilarity is that the polarity of the output signal is not inverted. Switching topology of the SEPIC converter is as shown in Fig. 6. During operating conditions in a stable state, the magnitude of the voltage across the inductor is zero, so the magnitude of the voltage on the capacitor C_1 is $V_{C1} = V_S$. When the condition for switch is closed then the diode is opened, the inductor L_1 is loaded from source V_S, and inductor L_2 fill C_1. During these conditions, no energy is delivered to the load. When the condition of the switch is open, diodes condition is closed, L_1 fill C_1 and L_2 transfer current to the load then the magnitude of the voltage passes 1, $V_{1\text{close}} = V_s$. The magnitude of the voltage on L_1 is V, $L_{1\text{open}} = V_0$ [5–7].

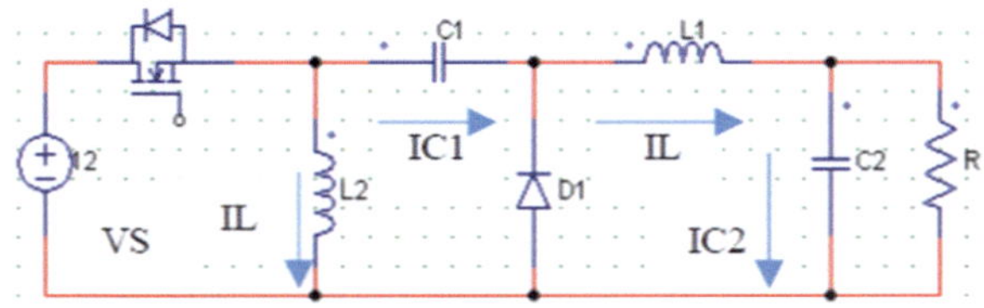

Fig. 7 Zeta converter switching topology

5.3 *ZETA Converter*

A Zeta converter is a fourth-order DC–DC converter made up of two inductors and two capacitors and capable of operating in either step-up or step-down mode. In ZETA DC–DC converter, there are two modes of operations. During first mode switch is made turn on and in the other mode switch is made to turn off. Zeta converter is shown in Fig. 7. It comprises of insulated-gate bipolar transistor (IGBT) as a switch, diode, two capacitor C_1 and $C_{2,}$ and two inductor L_1 and L_2 with a load *RL*. In the first operation mode, inductor L_1 and L_2 start charging. In the second operation mode, inductors (L_1, L_2) are in a condition to release the stored energy (discharging). The release of energy from L_1 charges the capacitor C_1 and the inductor L_2 transfer energy to the output circuit which is connected to the load [5–7].

6 DC–DC Converters: Comparison

Comparison of several DC–DC converters illustrates the input power signal of PV and output of the DC–DC converters circuit.

Comparison of results in Figs. 8, 9 and 10 show that maximum power point tracking signal occurred best in ZETA converter where level of ripple is low in current and voltage.

7 Conclusion

This paper shows the review work of previously used control techniques. Firstly, P&O, IPSO, and GWO algorithm are compared in which GWO algorithm shows better result as IPSO and GWO overcome the problem of local minima. Various DC–DC converters are also compared that are CUK, SEPIC, ZETA converters. The comparison of these converters shows that ZETA converter shows more effectiveness than other two converters.

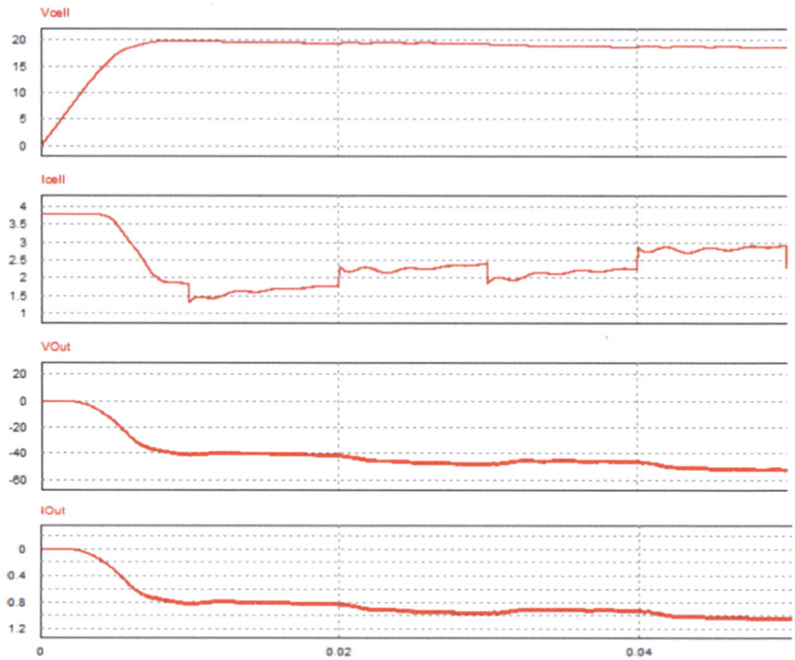

Fig. 8 CUK converter

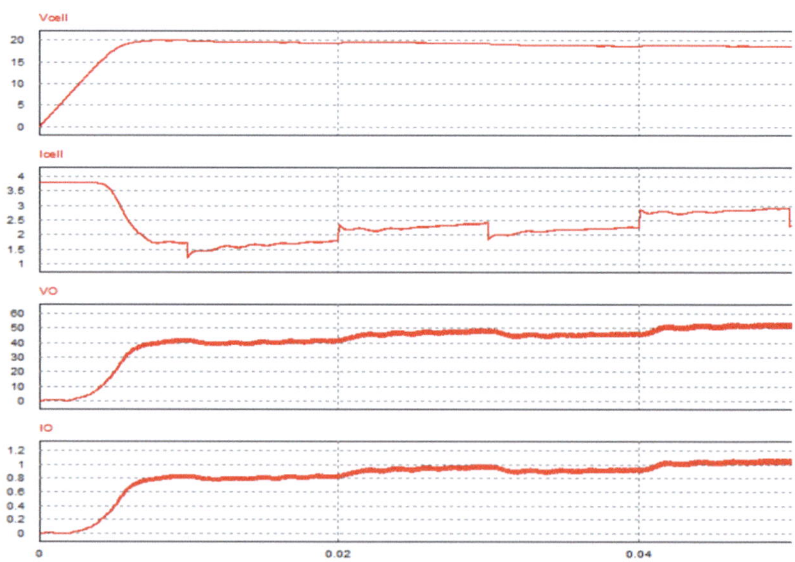

Fig. 9 SEPIC converter

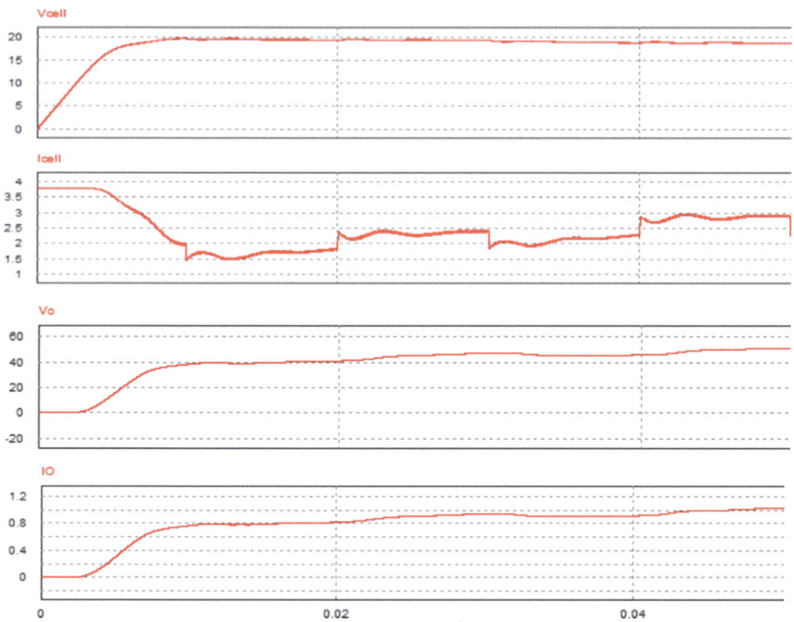

Fig. 10 ZETA converter

References

1. Alsadi, S., Alsayid, B.: Maximum power point tracking simulation for photovoltaic systems using perturb and observe algorithm. Int. J. Eng. Innov. Technol. (IJEIT) **2**(6) (2012). ISSN: 2277-3754
2. Mohanty, S., Subudhi, B., Ray, P.K.: A new MPPT design using grey wolf optimization technique for photovoltaic system under partial shading conditions. IEEE Trans. Sustain. Energy **7**(1), 1949–3029 (2016)
3. Mahmoud, Y., Abdelwahed, M., El-Saadany, E.F.: An enhanced MPPT method combining model-based and heuristic techniques. IEEE Trans. Sustain. Energy **7**(2), 1949–3029 (2016)
4. Yan, X., Wu, Q., Liu, H., Huang, W.: An improved particle swarm optimization algorithm and its application. IJCSI Int. J. Comput. Sci. Issues **10**(1), No 1 (2013). ISSN (Print): 1694-0784|ISSN (Online): 1694-0814
5. Soedibyo, M.A., Amri, B.: The comparative study of Buck-boost, Cuk, Sepic and Zeta converters for maximum power point tracking photovoltaic using P&O method. In: Proceedings of 2015 2nd International Conference on Information Technology, Computer and Electrical Engineering (ICITACEE), Indonesia, 978-1-4799-9863-0/15
6. Rajeswari, R.V., Geetha, A.: Comparison of Buck-boost and CUK converter control using fuzzy logic controller. Int. J. Innov. Res. Sci. Eng. Technol. **3**(3) (2014)
7. Bist, V., Singh, B.: A brushless DC motor drive with power factor correction using isolated zeta converter. IEEE Trans. Industr. Inf. **10**(4), 1551–3203 (2014)

Performance Analysis of PV Cell Using One Diode and Two Diode Model

Noorul Islam, A. Q. Ansari and L. Navinkumar Rao

Abstract The simulation and modeling of photovoltaic (PV) cell using different diode configurations are presented in the paper under different test conditions. The performance characteristics have been evaluated based on the variation in temperature, solar radiations, series resistance, shunt resistance, and ideality factor (IF) in the proposed model. The results are compared for maximum power, maximum voltage, and maximum current with respect to standard values. This paper also discusses the effects on the performance of PV cell due to variation of environmental and physical parameters with respect to standard test conditions.

Keywords Photovoltaic cell · One diode model (ODM) · Two diode model (TDM) · Insolation · Series resistance · Parallel resistance · Ideality factor (IF)

1 Introduction

Photovoltaic systems are designed to supply green power to end user by means of photovoltaic cells. Photovoltaic system consists of many components including solar panel, charge controller, inverter. The basic unit of PV panel is the PV cell. A standard PV cell produces energy less than 3 watts at approximately 0.6 V dc. It should be connected in different (e.g., series–parallel) arrangements to produce suffice solar power for heavy load applications. Figure 1 shows how PV cells are configured into modules, and how modules are connected as array.

N. Islam (✉)
Jamia Millia Islamia, New Delhi, India
e-mail: noorulislam@its.edu.in

A. Q. Ansari
Electrical Engineering, Jamia Millia Islamia, New Delhi, India
e-mail: aqansari62@gmail.com

L. Navinkumar Rao
Electrical Engineering, Nagpur Institute of Technology, Nagpur, India
e-mail: navinkamal@gmail.com

S. N. Singh et al. (eds.), *Advances in Energy and Power Systems*, Lecture Notes in Electrical Engineering 508, https://doi.org/10.1007/978-981-13-0662-4_5

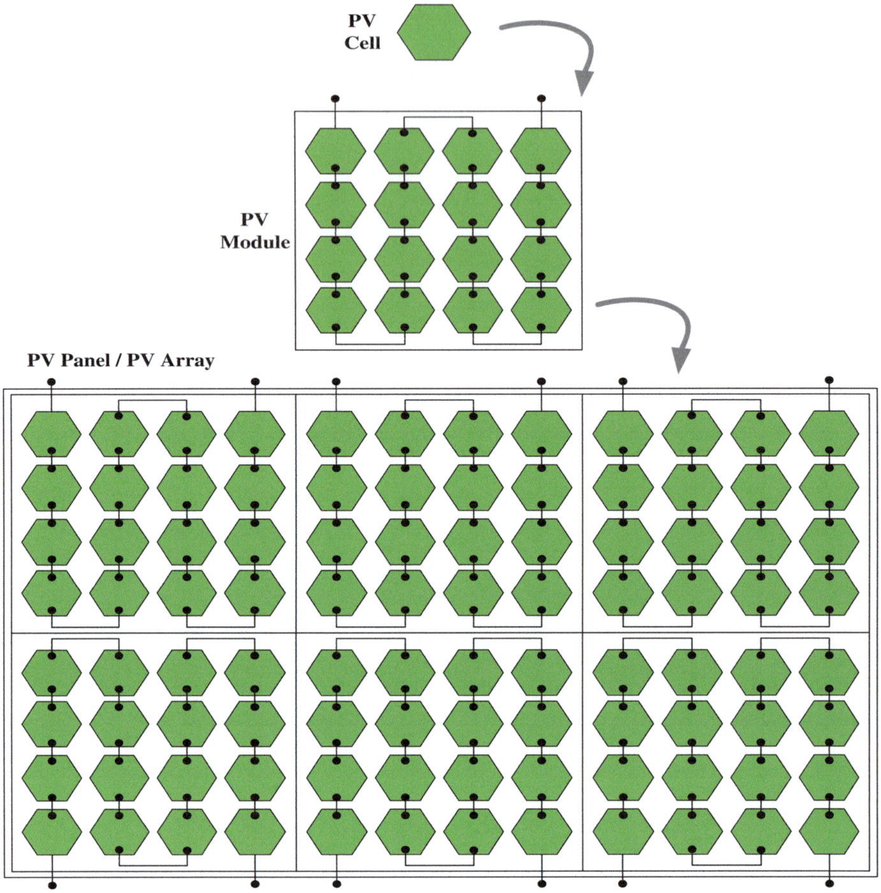

Fig. 1 PV cell, module, and panel models

PV cells are being used to generate electrical energy directly from sunlight. The electrical energy received from sunlight is utilized by individual capacity to big organization (fractional watts to 100 of Megawatts). It is emission-free energy which is helpful in reducing global warming. The running cost of solar energy is almost negligible as compared to other energy resources. Here, we have decided to analyze the performance of solar cell. There are two equivalent model presented in various journals, viz. ODM [1–4] and TDM [4–9].

The equivalent circuit model of PV cell is represented by a current source in parallel with diode (ODM or TDM) as shown in Fig. 2. It consists of a current source, a diode, parallel resistance, and series resistance. The mathematical equation of photovoltaic cell, including photocurrent source, diode, series resistor, and shunt resistor [1], is described by MATLAB Simulink (graphical programming environment, developed by Mathworks). The parameters extraction of nonlinear I–V equation by adjusting the curve at three points, open circuit, maximum power, and short circuit is discussed in [2]. The dependence of irradiance on the maximum power output of a PV module is reproduced by Kenneth et al. [3]. Author has proposed an accurate model for the methodical modification of parameters of ODM. Electrical performance of nanorod-based dye-sensitized solar cells (DSSCs) 1-D zinc oxide (ZnO) [4] were theoretically analyzed and experimental measurement using simple ODM is discussed by researchers. A comprehensive solar photovoltaic power electronic conversion system is simulated [5]. The electrical properties such as temperature, irradiance, resistance and their impacts on the power output can be identified using distributed modeling [6]. Controlled current source and S-function builder PV model is simulated in MATLAB by researcher [7]. The performance characteristics (I–V and P–V) of a solar panel, having a large number of different series and parallel connected modules [8], under partially shaded conditions is proposed by the researcher. Variation of maximum power with temperature and insolation levels was investigated [9]. The PV cell general model is developed and implemented [10] on simulation platform of Personal Computer Simulation Program with Integrated Circuit Emphasis (PSPICE). PV cell parameters extracted from derived analytical expressions [11–13] of solar cell diode model from experimental data. A better modeling approach for the TDM of photovoltaic module is presented [14–16]. There are many solutions proposed on PV cell parameters extraction. Analytical solutions for the rapid extraction have been proposed from experimental data [17] of different diode models. A photovoltaic module using one diode with five parameters capable to predict the PV panel characteristics in different temperature and solar insolation conditions built and tested [18]. A simple experimental-arrangement technique [19] for solar cell was proposed. In this technique, dark current is measured with calibrated standard resistance for calibration of series resistance. Simulink-based simulation study of solar cell was carried out [20] and complete performance study was performed for varying conditions of solar irradiation, temperature, variable diode model parameters, series and parallel resistance.

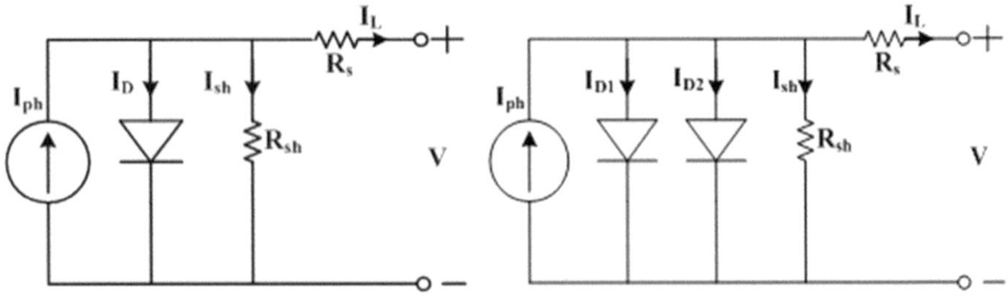

(b) ODM **(b)** TDM

Fig. 2 **a** ODM. **b** TDM

2 Photovoltaic Cell

PV cell, PV module, and PV panel are shown in Fig. 1. The general equivalent circuit diagrams of PV cell are shown in Fig. 2a for ODM and Fig. 2b for TDM. It consists of a current source, parallel diode, a shunt resistor providing leakage current path, and a series resistor describing an internal resistance to the current flow [1, 10]. The modeling of the PV cell is developed based on Eqs. (1)–(3).

3 Modeling of Photovoltaic Cell

3.1 Odm

The PV cell is the building block of a photovoltaic module and this is responsible for the transforming the sunlight or photons directly into electric energy. A semiconductor material is used in PV cell, whose electrical characteristics differ similar to a diode. The diode current in PV cell is given by the Shockley Eq. (1).

$$I_D = I_0\left[e^{\frac{qV}{AKT}} - 1\right] \quad (1)$$

where:

I_D Diode dark current (A),
I_0 Diode reverse bias saturation current (A),
V Voltage across PV cell (V),
q Electron charge (1.6×10^{-19} C),
A Diode IF (1.0–2.0),
K Boltzmann's constant (1.38×10^{-23} J/K),
T Solar cell temperature (1.0–2.0),

From equivalent circuit diagram as shown in Fig. 2a,

$$I_L = I_{\text{ph}} - I_D - I_{\text{sh}} \tag{2}$$

$$\text{Where,} \quad I_{\text{sh}} = \frac{V - IR_s}{R_{\text{sh}}} \quad \text{and} \tag{3}$$

I_L is the PV cell load current, I_{ph} is the photocurrent (the current generated by the incident light, directly proportional to the sun insolation) and leakage current I_{sh} given by Eq. (2).

Combining Eqs. (1)–(3), we get,

$$I_L = I_{\text{ph}} - I_0\left[\text{e}\frac{q(V + IR_s)}{AKT} - 1\right] - \frac{V + IR_s}{R_{\text{sh}}} \tag{4}$$

Therefore, in ODM, there are five parameters of interest, i.e., I_{ph}, I_0, A, R_s, R_{sh}, which effect the performance of solar cell.

3.2 *TDM*

The equivalent circuit diagram of TDM of PV cell is shown in Fig. 2b, with considering series and parallel resistance. The resultant current of TDM is given by the Eq. (5). Where I_{D1}, I_{D2} is the normal diode current. The PV cell load current I_L is given by,

$$I_L = I_{\text{ph}} - I_{D1} - I_{D1} - I_{\text{sh}} \tag{5}$$

$$\text{Where,} \quad I_{D1} = I_{01}\left[\text{e}\frac{qV}{A_1KT} - 1\right] \tag{6}$$

$$I_{D2} = I_{02}\left[\text{e}\frac{qV}{A_2KT} - 1\right] \tag{7}$$

Combining Eqs. (3), (5)–(7), we get,

$$I_L = I_{\text{ph}} - I_{01}\left[\text{e}\frac{q(V + IR_s)}{A_1KT} - 1\right] - I_{01}\left[\text{e}\frac{q(V + IR_s)}{A_2KT} - 1\right] - \frac{V + IR_s}{R_{\text{sh}}} \tag{8}$$

From Eq. (5), we find that there are seven parameters interest, i.e., I_{ph}, I_{01}, I_{02}, A_1, A_2, R_s, R_{sh}, which effect the performance of solar PV cell.

Therefore, the comparison of characteristics for changing various parameters in ODM and TDM is presented in this paper.

Table 1 Parameters for the ODM and TDM

CIS-thin film SOLARWORLD ST40 (peak power: 40 W)		
	ODM	TDM
V_{mp} (V)	16.66	16.66
I_{mp} (A)	2.41	2.41
V_{oc} (V)	23.3	23.3
I_{sc} (A)	2.68	2.68
R_s (Ω)	0.534	0.717
R_{sh} (Ω)	2940	1590

Table 2 Comparison of various parameters of ODM and TDM at standard test conditions

Parameters	ODM ($A = 1.3$)	TDM ($A_1 = 1.3, A_2 = 2$)	Solerex_40 model (manufacturer specification)
Maximum power (W_P) (W)	42.3	39.8	40
Maximum voltage (V_{MPP}) (V)	16.7	17.3	16.66
Maximum current (I_{MPP}) (A)	2.53	2.31	2.41
Open circuit voltage (V_{oc}) (V)	23.3	23.3	23.3
Short circuit current (I_{sc}) (A)	2.68	2.68	2.68

4 Modeling and Results

In order to compare photovoltaic models, and to analyze the effect of changing different parameters, a CIS-Thin film SOLARWORLD ST40 solar photovoltaic module of 40 W is considered. The manufacturer specifications of photovoltaic module are given in Table 1. These parameters are used in MATLAB programming to predict the performance of PV cell. Table 2 shows that performance of ODM and TDM under standard test conditions.

4.1 Effect of Temperature

The photocurrent and diode reverse saturation current is given as;

$$I_{ph} = [I_{sc} + K_I(T - 298)] \times \frac{B_x}{1000} \tag{9}$$

(a) ODM

(b) TDM

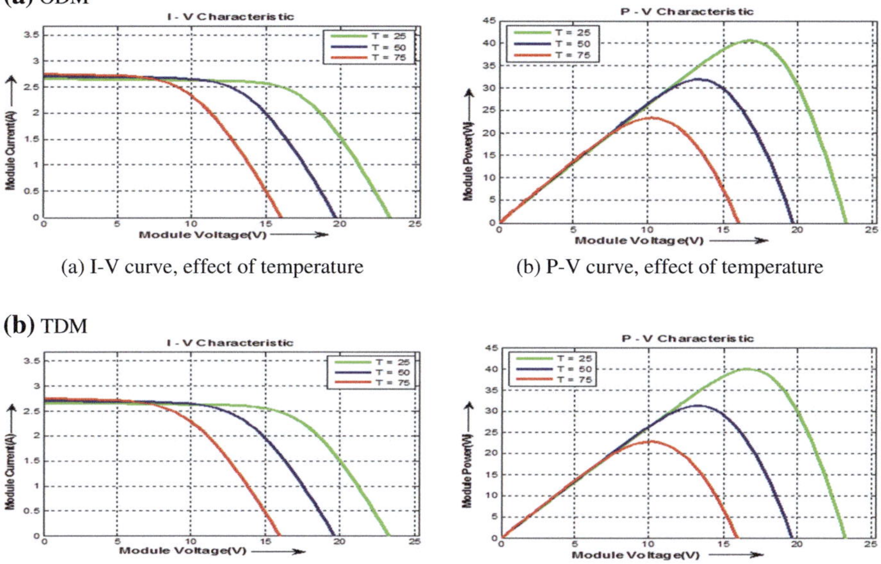

(a) I-V curve, effect of temperature

(b) P-V curve, effect of temperature

(c) I-V curve, effect of temperature

(d) P-V curve, effect of temperature

Fig. 3 **a** I-V curve, effect of temperature. **b** P-V curve, effect of temperature. **c** I-V curve, effect of temperature. **d** P-V curve, effect of temperature

$$I_{\text{rs}} = I_0 \left[\frac{T}{T_r} \right]^3 \times \mathrm{e} \frac{qE_g}{AK} \left[\frac{1}{T_r} - \frac{1}{T} \right] \tag{10}$$

where K_I is the short circuit current temperature constant of PV cell and B_x is the solar insolation in Watt/m^2.

From the above Eqs. (9) and (10), it is understood that the diode reverse saturation current (I_0) and photocurrent of PV cell depends on the temperature variation linearly and quadratically, respectively. Hence, it can be concluded that the change of I_0 is more dominant as compared to I_{ph}. The output I–V and P–V characteristics of PV model for different input temperatures [9, 10] are shown in Fig. 3.

4.2 Effect of Insolation

The solar cell photocurrent directly proportional to the solar insolation [7], i.e.,

$$I_{\text{ph1}} = \frac{B_1}{B_0} \times I_{\text{ph0}} \tag{11}$$

where I_{ph1} is solar cell photocurrent at solar insolation B_1.

(a) ODM

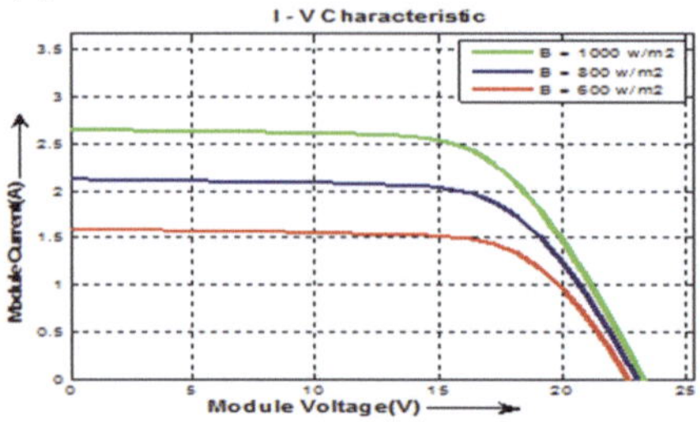

(a) I-V curve, effect of solar insolation

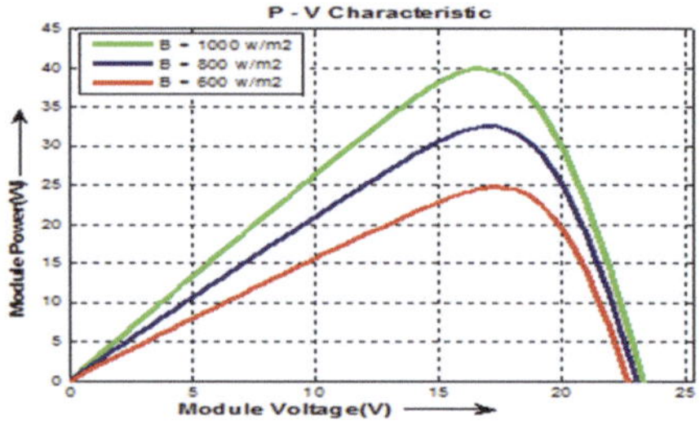

(b) P-V curve, effect of solar insolation

(b) TDM

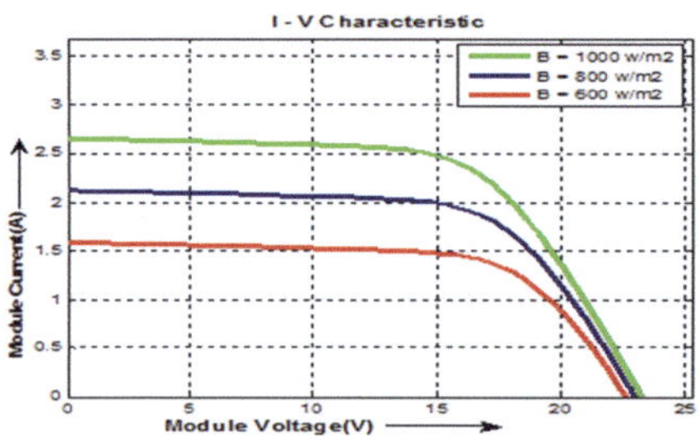

(c) I-V curve, effect of solar insolation

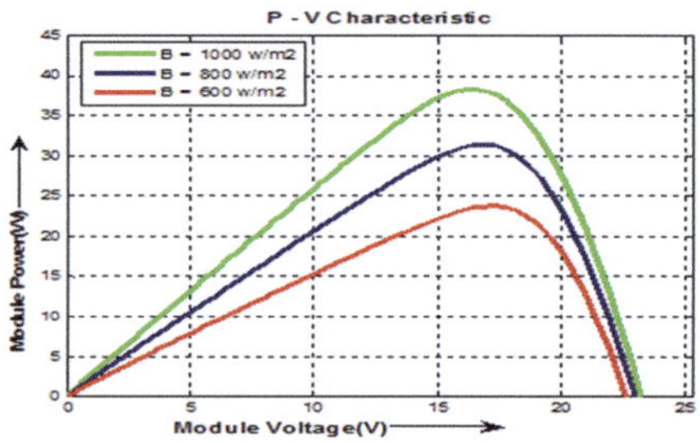

(d) P-V curve, effect of solar insolation

Fig. 4 **a** I-V curve, effect of solar insolation. **b** P-V curve, effect of solar insolation. **c** I-V curve, effect of solar insolation. **d** P-V curve, effect of solar insolation

The effects of solar insolation on photocurrent are shown in Fig. 4.

4.3 Effect of Series Resistance

The series resistance of a PV cell is very low and it may be neglected for small power applications [20], but in large area commercial solar array, series resistance of a PV cell affects the fill factor losses [21]. Therefore, an accurate measurement is necessary for measuring losses [10, 19, 20]. The performance characteristics of PV cell with the variation of series resistance are shown in Fig. 5.

4.4 Effect of Parallel Resistance

The parallel resistance present in PV cell is normally due to manufacturing defects and not due to poor cell design. Power losses in PV cells are due to low parallel

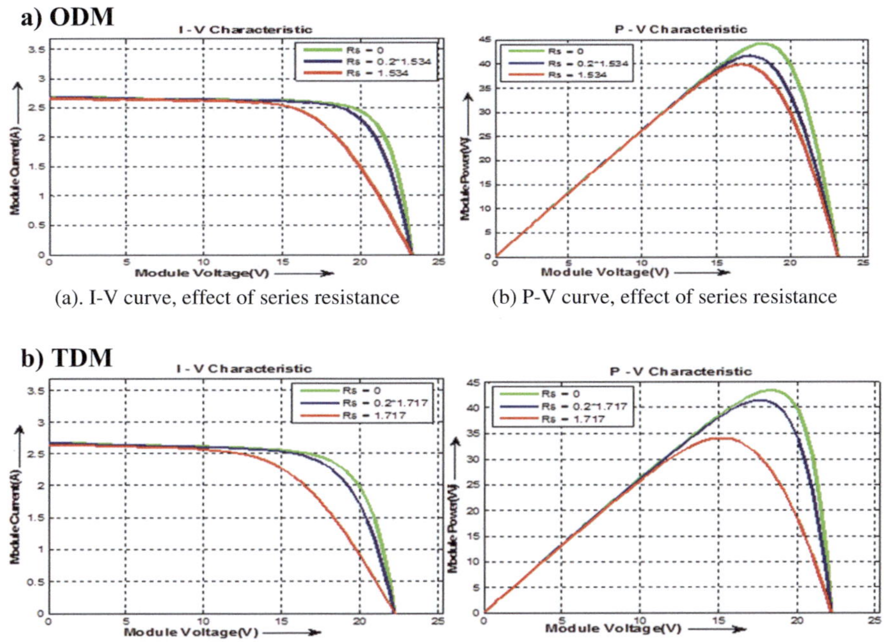

(a). I-V curve, effect of series resistance

(b) P-V curve, effect of series resistance

(c) I-V curve, effect of series resistance

(d) I-V curve, effect of series resistance

Fig. 5 **a** I-V curve, effect of series resistance. **b** P-V curve, effect of series resistance. **c** I-V curve, effect of series resistance. **d** I-V curve, effect of series resistance

resistance. This resistance provides an alternate current path to the photogenerated current. This diversion of current reduces the PV cell junction current and reduces the PV cell voltage. The effect of a parallel resistance is particularly critical at low insolation levels; at low insolation level, the photogenerated current is also low. Therefore, the loss of this leakage current to the alternate path has a bigger impact on the performance of PV cell. In addition, when the effective resistance of PV cell is high at lower voltage, the impact of parallel resistance is high [10]. The effects of parallel resistance are shown in Fig. 6.

4.5 *Effect of Ideality Factor (iF)*

Here in the proposed model, the IF of ODM in PV cell equation is assumed as a constant value. Since, IF depends on the voltage across the PV cell, therefore, at higher voltage, IF of diode is close to 1. Also it has been observed that, at higher voltage the recombination of electrons and holes in the semiconductor device is influenced by the type of surface and the bulk regions [1, 10, 13]. However, at lower voltages, as the recombination of electrons and holes in the junction dominates.

a) ODM

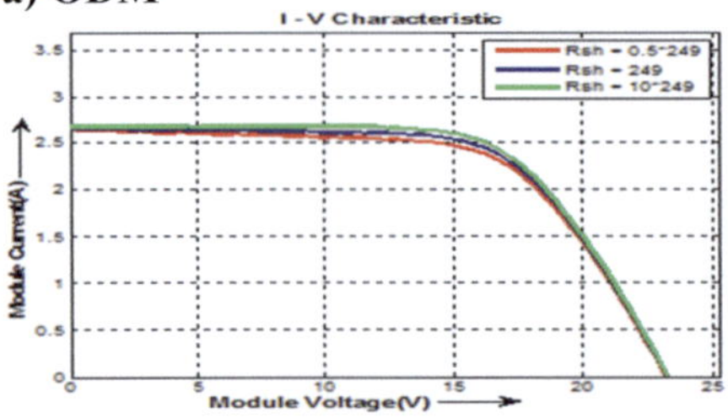

(a). I-V curve, effect of shunt resistance

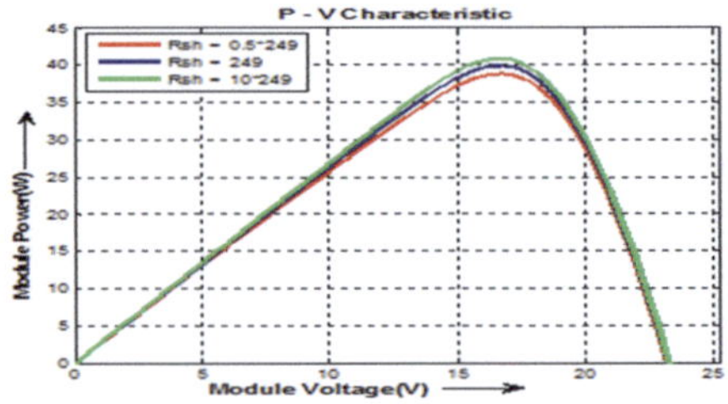

(b) P-V curve, effect of shunt resistance

b) TDM

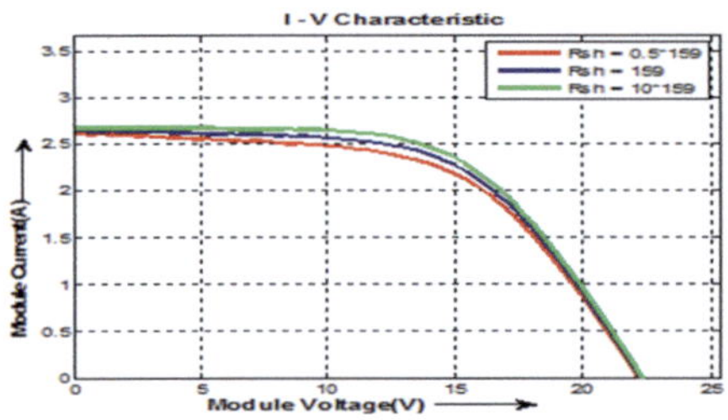

(c) I-V curve, effect of shunt resistance

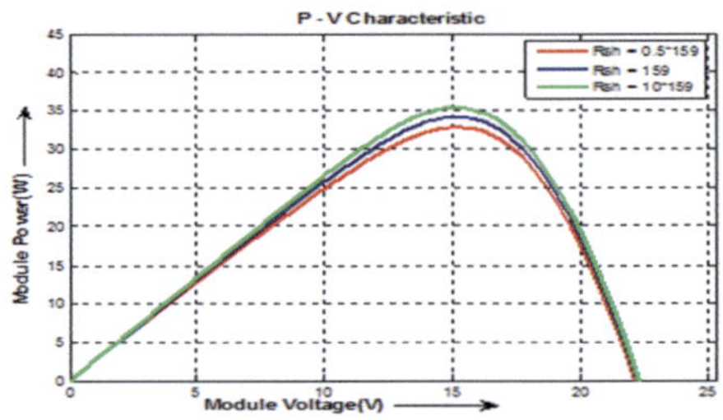

(d) P-V curve, effect of shunt resistance

Fig. 6 **a** I-V curve, effect of shunt resistance. **b** P-V curve, effect of shunt resistance. **c** I-V curve, effect of shunt resistance. **d** P-V curve, effect of shunt resistance

Therefore, IF in TDM become nearly 2. The performance characteristics of ODM and TDM with variation of IF are given in Fig. 7.

5 Conclusion

Above simulation results have shown the performance characteristics of PV model for different insolation, temperature, series resistance, shunt resistance, and IF. The performance characteristics have been compared between ODM and TDM, for peak power, peak voltage, peak current, under standard test conditions using MATLAB Simulink. The TDM PV cell poses better performance as compared to ODM in terms of peak power, peak voltage, peak current close to manufacturer specifications.

a) ODM

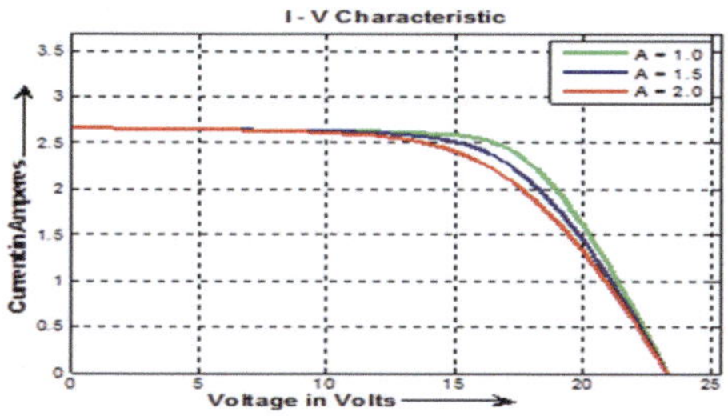

(a). I-V curve, effect of IF

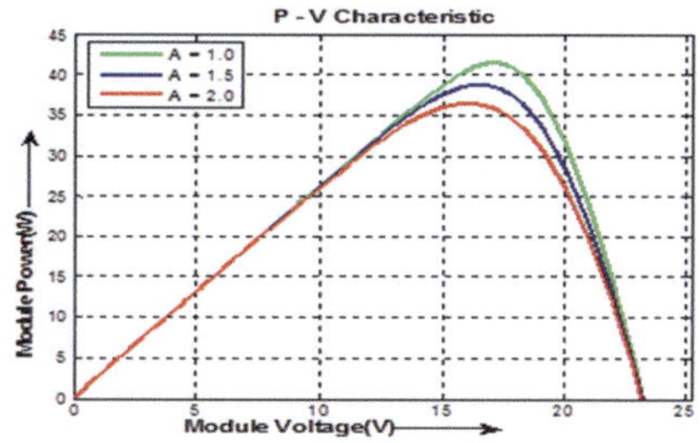

(b) P-V curve, effect of IF

b) TDM

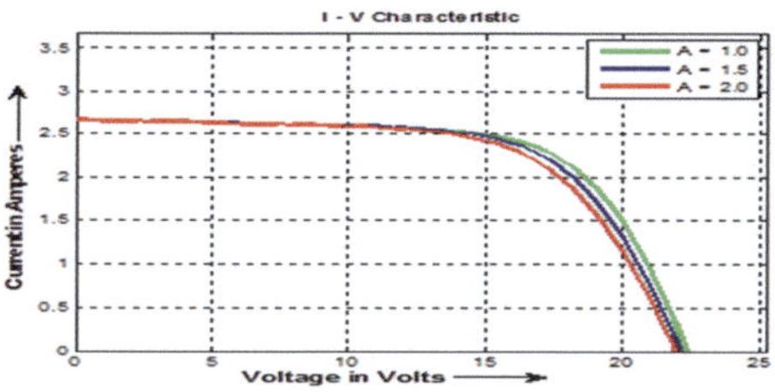

(c) I-V curve, effect of IF

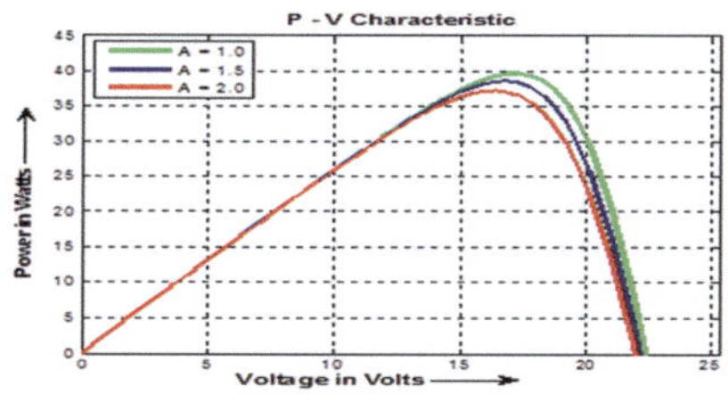

(d) P-V curve, effect of IF

Fig. 7 **a** I-V curve, effect of IF. **b** P-V curve, effect of IF. **c** I-V curve, effect of IF. **d** P-V curve, effect of IF

References

1. Tarak, S., Mounir, B., Adel, G., Ahmed, M.: MATLAB/Simulink based modelling of solar photovoltaic cell. Int. J. Renew. Energy Res. **2**(2), 213–218 (2012)
2. Villalva, M.G., Gazoli, J.R.: Filho ER (2009) Comprehensive approach to modeling and simulation of photovoltaic arrays. IEEE Trans. Power Electron. **24**(5), 1198–1208 (2009)
3. Kenneth, J.S., Thomas, R., Clifford, W.H.: Modeling the irradiance and temperature dependence of photovoltaic modules in PVsyst. IEEE J. Photovolt. **5**(1), 152–158 (2015)
4. Kyaw, H.H., Bora, T., Joydeep, D.: One-diode model equivalent circuit analysis for ZnO nanorod-based dye-sensitized solar cells: effects of annealing and active area. IEEE Trans. Nano-Technol. **11**(4), 763–768 (2012)
5. Villalva, M., Gazoli, J.: Comprehensive approach to modeling and simulation of photovoltaic arrays. IEEE Trans. Power Electron. **24**(5), 1198–1207 (2009)
6. Wu, X., Bliss, M., Sinha, A., Betts, T., Gupta, R., Gottschalg, R.: Renew. Gener. Power Gener. **8**(5), 459–466 (2014)
7. Ding, K., Bian, X.G., Liu, H.H., Peng, T.: A MATLAB-Simulink-based PV module model and its application under conditions of non-uniform irradiance. IEEE Trans. Energy Convers. **27**(4), 864–872 (2012)
8. Patel, H., Agarwal, V.: MATLAB-based modeling to study the effects of partial shading on PV array characteristics. IEEE Trans. Energy Convers. **23**(1), 302–310 (2008)
9. Walker, G.: Evaluating MPPT converter topologies using a MATLAB PV model. J. Electric. Electron. Eng. Australia **21**(1), 49–56 (2001)

10. Gow, J., Manning, C.: Development of a photovoltaic array model for use in power-electronics simulation studies. In: IEE Proceedings, Electric Power Applications, pp. 193–200, March 1999)
11. Phang, J.C.H., Chan, D.S.H., Phillips, J.R.: Accurate analytical method for the extraction of solar cell model parameters. Electron. Lett. **20**, 406–408 (1984)
12. Chan, D.S.H., Phillips, J.R., Phang, J.C.H.: A comparative study of extraction methods for solar cell model parameters. Solid State Electron. **29**, 329–337 (1986)
13. Chan, D.S.H., Phang, J.C.H.: Analytical methods for the extraction of solar-cell single- and double-diode model parameters from I-V characteristics. IEEE Trans. Electron Dev. **ED-34**(2), 286–293 (1987)
14. Elbaset, A.A., Ali, H., Abd-El Sattar, M.: Novel seven-parameter model for photovoltaic modules. Solar Energy Mat. Solar Cells **130**, 442–455 (2014)
15. Ishaque, K., Salam, Z., et al.: Simple, fast and accurate two-diode model for photovoltaic modules. Solar Energy Mat. Solar Cells **95**(2), 586–594 (2011)
16. Ishaque, K., Salam, Z., Taheri, H.: Accurate MATLAB simulink PV system simulator based on a two-diode model. Power Electron. **11**(2), 179–187 (2011)
17. Kassis, A., Saad, M.: Analysis of multi-crystalline silicon solar cells at low illumination levels using a modified two-diode model. Solar Energy Mat. Solar Cells **94**, 2108–2112 (2010)
18. Chenni, R., Makhlouf, M., Kerbache, T., Bouzid, A.: A detailed modeling method for photovoltaic cells. Energy **32**, 1724–1730 (2007)
19. Sera D, Teodorescu R, Rodriguez P (2007) PV panel model based on datasheet values. In: IEEE Industrial Electronics Conference, July 2007
20. Cabestany, J., Castaner, L.: A simple solar cell series resistance measurement method. Rev. de Phys. Appliquee **18**(9), 565–567 (1983)
21. Masri, S., Chan, P.W.: Design and development of a dc-dc Boost converter with constant output voltage. In: IEEE, International Conference on Intelligent and Advanced Systems (ICIAS), pp. 1–4, 15–17 June 2010

Part II
Power Management

Active Power Regulation by MPC based Flywheel Energy Storage System

Aasim, S. N. Singh and Abheejeet Mohapatra

Abstract The flywheel is used for power regulation in wind energy system (WES) due to stochastic nature of wind. The power exchange is equivalent to speed control of connected machine above and below its rated speed. Traditionally, doubly fed induction machine (DFIM) coupled to a flywheel uses a proportional–integral (PI) control for both outer speed-control loop and inner current-control loop. This paper proposes modal predictive control (MPC)-based inner current-control loop, which reduces the speed overshoot for a given speed change as compared to PI-based inner control loop, while settling time for speed control through MPC is comparable to PI-based control. Active power regulation due to 500 kW, 100 kg m^2 flywheel energy system integrated with 1.5 MW type-3 wind turbine generator is considered for variable wind condition in MATLAB/SIMULINK.

Keywords Doubly fed induction machine · Flywheel · Speed control · Modal predictive control

1 Introduction

High penetration of wind energy in a conventional power grid is encouraged due to the gradual depletion of fossil fuel and increased concerns about global warming. Wind speed, being stochastic in nature, is making energy extracted from it non-dispatchable in nature. Wind energy system (WES) is now coupled with energy storage devices such as flywheel, battery, supercapacitor energy storage, superconducting magnetic storage system [1]. A flywheel requires less maintenance, quick acting, mechanically robust, high energy efficient, least environmental pollutant. It quickly reacts

Aasim (✉) · S. N. Singh · A. Mohapatra
Department of Electrical Engineering, Indian Institute of Technology Kanpur, Kanpur, India
e-mail: aasim@iitk.ac.in

S. N. Singh
e-mail: snsingh@iitk.ac.in

A. Mohapatra
e-mail: abheem@iitk.ac.in

S. N. Singh et al. (eds.), *Advances in Energy and Power Systems*, Lecture Notes in Electrical Engineering 508, https://doi.org/10.1007/978-981-13-0662-4_6

to variation in electrical output of wind energy system, thereby delivering smoother power to grid [2].

A typical flywheel energy storage system (FESS) comprises of machine, bearing, and power electronics converter-based interface (PECI). The machine, which acts as an electromechanical interface of FESS, may be induction machine (IM), doubly fed induction machine (DFIM), synchronous machine (SM), permanent magnet synchronous machine (PMSM), while PECI comprises of a variety of different converter configurations [2]. The bearing of FESS has evolved from mechanical bearing to magnetic bearing. However, in spite of long life, fast response, low losses in magnetic bearing, FESS still uses auxiliary mechanical bearing in case of fault/overloading in magnetic bearing [2].

Integrated operation of FESS with WES is an explored area. Fuzzy supervised power reference to FESS is studied in [3], power reference to FESS using low-pass filter in [4], and moving average method in [5]. FESS is controlled for fault ride through support in [6]. Energy management of WES through FESS is dealt in [7]. Control strategies of FESS can be categorized under flux-oriented control (FOC) and direct torque control (DTC). While the majority of research focuses on FOC, [4] uses DTC. FOC is basically two-level control. The outer loop which has a large time constant is voltage-control loop, while the inner loop having large bandwidth is current-control loop. It generally uses modulator for generating gate signal for switches of the converter. DTC, which is single-level control, uses hysteresis controller.

Regulation of active power is achieved through energy exchange between WES and FESS. During high-wind speed condition, a portion of energy converted by WES is stored in FESS which is reflected as increase in flywheel speed, whereas during low-wind speed condition, the stored energy in flywheel is released and a corresponding decrease in flywheel speed is observed. Accordingly, this active power exchange between WES and FESS can be translated as speed control of FESS. Speed control of FESS with less overshoot and small settling time will give an efficient energy exchange with WES.

This paper proposes modal predictive control (MPC) of FESS for active power regulation of WES. MPC is a general approach, wherein modal of the system is used to predict the future state considering the present state of the system. Initially, MPC was used for slow time-varying system, but nowadays, it finds its use in electrical system due to fast computational feature of microprocessor. The outer loop of FESS is a PI-based controller, while the inner loop is MPC-based. This proposed control is then compared with a PI-based inner control loop for the same step change in speed command. A comparative study for speed overshot by PI–MPC and PI–PI control is also made. FESS is then integrated with WES as shown in Fig. 1, and active power regulation under varying wind condition is studied.

The next section deals with dynamic modal of DFIM. Vector control of DFIM is explained in Sect. 3. A brief introduction to MPC and its grid-side voltage-source converter (VSC) and rotor-side VSC is given in Sect. 4. Simulation results are shown in Sect. 5, and conclusion of the work is presented in Sect. 6.

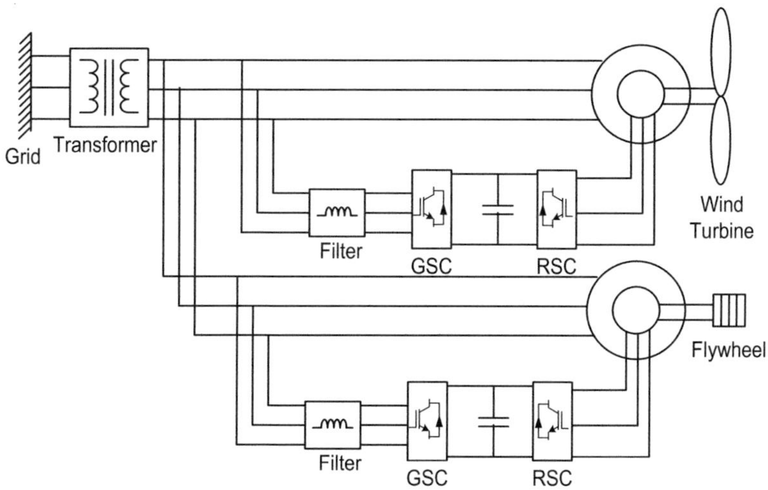

Fig. 1 FESS-WES integrated system under study

2 DFIM Dynamic Model

The electrical dynamical model of DFIM in a synchronously rotating (*dq*) reference frame is given by the following equations [8]:

$$\left.\begin{aligned} v_{ds} &= R_s i_{ds} + \frac{d\psi_{ds}}{d_t} - \omega_s \psi_{qs} \\ v_{qs} &= R_s i_{qs} + \frac{d\psi_{qs}}{d_t} + \omega_s \psi_{ds} \end{aligned}\right\} \quad (1)$$

$$\left.\begin{aligned} v_{dr} &= R_r i_{dr} + \frac{d\psi_{dr}}{d_t} - (\omega_s - \omega_r)\psi_{qr} \\ v_{qr} &= R_r i_{qr} + \frac{d\psi_{qr}}{d_t} + (\omega_s - \omega_r)\psi_{dr} \end{aligned}\right\} \quad (2)$$

$$\left.\begin{aligned} \psi_{ds} &= L_s i_{ds} + L_m i_{dr} \\ \psi_{qs} &= L_s i_{qs} + L_m i_{qr} \end{aligned}\right\} \quad (3)$$

$$\left.\begin{aligned} \psi_{dr} &= L_r i_{dr} + L_m i_{ds} \\ \psi_{qr} &= L_r i_{qr} + L_m i_{qs} \end{aligned}\right\} \quad (4)$$

where

$$\left.\begin{aligned} L_s &= L_{ls} + L_m \\ L_r &= L_{lr} + L_m \end{aligned}\right\} \quad (5)$$

where $v_{ds}, v_{qs}, i_{ds}, i_{qs}, \psi_{ds}, \psi_{qs}$ are voltages, currents, and flux linkages of the stator in d- and q-axes. Similarly, $v_{dr}, v_{qr}, i_{dr}, i_{qr}, \psi_{dr}, \psi_{qr}$ are corresponding quantities for the rotor of DFIM. L_s, L_r, L_m are stator, rotor, and mutual inductances. L_{ls}, L_{lr} are leakage inductance of stator and rotor; ω_s is the electrical speed of the synchronously rotating reference frame, while ω_r is the electrical angular speed of the machine rotor.

The mechanical model of the DFIM is given by:

$$\left.\begin{aligned} J\frac{\mathrm{d}\Omega_m}{\mathrm{d}t} &= T_e - T_l - k_f\Omega_m \\ \frac{\mathrm{d}\theta_m}{\mathrm{d}t} &= \Omega_m \end{aligned}\right\} \tag{6}$$

where J is the combined inertia of the rotor and the load; Ω_m is the mechanical speed of the rotor; T_e is the electromagnetic torque generated; T_l is the external load torque; k_f is the combined rotor and load viscous friction coefficient.

The active and reactive powers of the stator DFIM can be expressed as:

$$\left.\begin{aligned} P_s &= v_{ds}i_{ds} + v_{qs}i_{qs} \\ Q_s &= v_{qs}i_{ds} - v_{ds}i_{qs} \end{aligned}\right\} \tag{7}$$

where, for rotor side, it is

$$\left.\begin{aligned} P_r &= v_{dr}i_{dr} + v_{qr}i_{qr} \\ Q_r &= v_{qr}i_{dr} - v_{dr}i_{qr} \end{aligned}\right\} \tag{8}$$

The electromagnetic torque is given by:

$$T_e = \frac{3}{2}p\frac{L_m}{L_s}\left(\psi_{qs}i_{dr} - \psi_{ds}i_{qr}\right) \tag{9}$$

3 Vector Control of DFIM

Vector control scheme, which had been developed for back-to-back converter-based speed control of DFIM, is used here. This control is decoupled along the real and reactive components of the input current of DFIM, where the machine is forced to behave dynamically as a DC machine through feedback control.

3.1 Grid-Side VSC (GSC)

Vector control of GSC is achieved through orientation of the reference frame along the grid voltage vector. This results in decoupled control of active and reactive powers flowing between the grid and GSC. Input current to GSC is regulated, with d-axis current $\left(i_{\text{conv},d}\right)$ regulating the DC-link voltage and q-axis current $\left(i_{\text{conv},q}\right)$ regulating the reactive power exchanged between the grid and GSC.

The dq-transformation equations of the voltage balance across inductor with respect to synchronously rotating reference frame rotating at ω_s are:

$$\left.\begin{aligned} v_{\text{grid},d} &= R_f i_{\text{conv},d} + L_f \frac{\mathrm{d}i_{\text{conv},d}}{\mathrm{d}t} - \omega_s L_f i_{\text{conv},q} + v_{\text{conv},d} \\ v_{\text{grid},q} &= R_f i_{\text{conv},q} + L_f \frac{\mathrm{d}i_{\text{conv},q}}{\mathrm{d}t} + \omega_s L_f i_{\text{conv},d} + v_{\text{conv},q} \end{aligned}\right\} \tag{10}$$

where R_f and L_f are the resistance and inductance of the filter, $v_{\text{grid},d}$ and $v_{\text{grid},q}$ are the d-axis and q-axis voltage of the grid, and $v_{\text{conv},d}$ and $v_{\text{conv},q}$ are the d-axis and q-axis voltage of the GSC, respectively.

DC-link voltage is controlled through PI-based controller whose output is the d-axis GSC current, whereas controlling for desired reactive power exchange by GSC gives the required q-axis GSC current.

3.2 Rotor-Side VSC (RSC)

Vector control of RSC is achieved through orientation of the reference frame along the stator flux vector. This results in decoupled control with active power directly controlled by q-axis rotor current $\left(i_{\text{rotor},q}\right)$ and reactive power controlled by d-axis rotor current $\left(i_{\text{rotor},d}\right)$. The outer loop, which tracks the reference speed, is controlled by a PI-based controller generating the reference current.

The dq-transformation equations of the voltage applied to the rotor circuit of DFIM with respect to stator flux vector rotating at ω_s are:

$$\left.\begin{aligned} v_{\text{rotor},d} &= R_r i_{\text{rotor},d} + \sigma L_r \frac{\mathrm{d}i_{\text{rotor},d}}{\mathrm{d}t} - (\omega_s - \omega_r)\sigma L_r i_{\text{rotor},q} \\ v_{\text{rotor},q} &= R_r i_{\text{rotor},q} + \sigma L_r \frac{\mathrm{d}i_{\text{rotor},q}}{\mathrm{d}t} + (\omega_s - \omega_r)\left(L_{\text{mm}} i_{\text{ms}} + \sigma L_r i_{\text{rotor},d}\right) \end{aligned}\right\} \tag{11}$$

where $v_{\text{rotor},d}$ and $v_{\text{rotor},q}$ are the d-axis and q-axis voltages of the RSC, $\sigma = \left(1 - \frac{L_m^2}{L_s L_r}\right)$ is the leakage coefficient; and $L_{mm} = \frac{L_m^2}{L_s}$, $i_{ms} = \frac{\psi_s}{L_s}$ is the magnetizing current of DFIM.

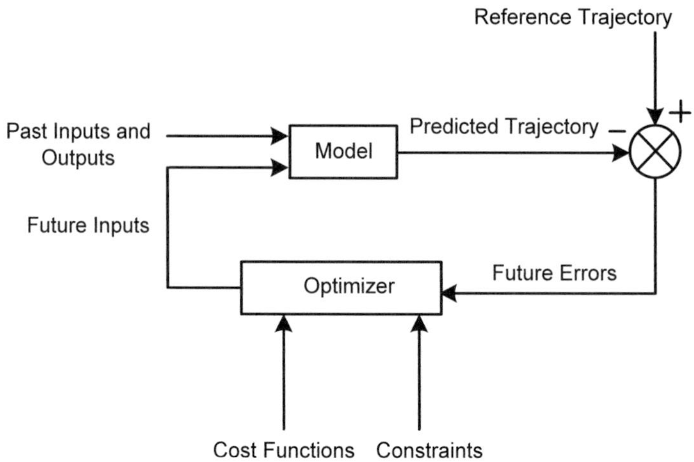

Fig. 2 Basic structure of MPC

4 Modal Predictive Control

Modal predictive control (MPC), which depends on system dynamical model, is an optimal control strategy. Recent development in processing power of microprocessors had resulted in MPC implementation in systems with a very small time constant. The basic of MPC can be understood through Fig. 2 [9].

The innermost part of MPC is the model of the plant to be controlled, which is used to predict the future values. The future values of the plant are predicted up to the prediction horizon p. These predicted values are then compared with the reference values, and an optimization algorithm determines the most optimal state of the plant under specified cost function and the system restriction.

As model of the system is continuous with respect to time, for a system of lower order, these continuous-time models are discretized by forward Euler method.

$$\frac{\mathrm{d}x}{\mathrm{d}t} = \frac{x(k+1) - x(k)}{T_s} \tag{12}$$

where T_s is the sample time of discretization process.

The discrete-time equation is then used to predict the state x at time instant $t = (k+1)$ using the value of the state at a time instant $t = k$.

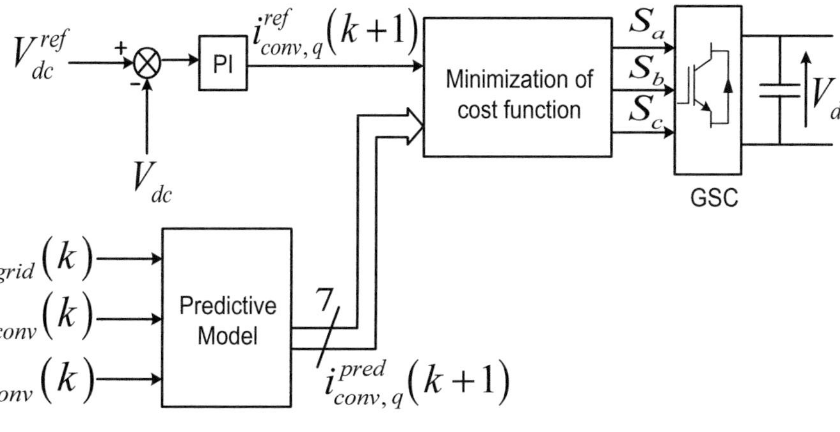

Fig. 3 MPC of grid-side VSC

4.1 MPC of Grid-Side VSC

Equation (10), which is a first-order differential equation, is discretized through forward Euler method to obtain the following discrete-time model of GSC:

$$\left.\begin{aligned} v_{\mathrm{grid},d}(k) &= R_f i_{\mathrm{conv},d}(k) + \frac{L_f}{T_s}\left[i_{\mathrm{conv},d}(k+1) - i_{\mathrm{conv},d}(k)\right] - \omega_s(k) L_f i_{\mathrm{conv},q}(k) + v_{\mathrm{conv},d}(k) \\ v_{\mathrm{grid},q}(k) &= R_f i_{\mathrm{conv},q}(k) + \frac{L_f}{T_s}\left[i_{\mathrm{conv},q}(k+1) - i_{\mathrm{conv},q}(k)\right] + \omega_s(k) L_f i_{\mathrm{conv},d}(k) + v_{\mathrm{conv},q}(k) \end{aligned}\right\} \tag{13}$$

The prediction of the current for $(k + 1)$ time instant is obtained from (13) as under:

$$\left.\begin{aligned} i_{\mathrm{conv},d}(k+1) &= \left[1 - \tfrac{R_f T_s}{L_f}\right] i_{\mathrm{conv},d}(k) + \omega_s(k) T_s i_{\mathrm{conv},q}(k) + \tfrac{T_s}{L_f}\left[v_{\mathrm{grid},d}(k) - v_{\mathrm{conv},d}(k)\right] \\ i_{\mathrm{conv},q}(k+1) &= \left[1 - \tfrac{R_f T_s}{L_f}\right] i_{\mathrm{conv},q}(k) - \omega_s(k) T_s i_{\mathrm{conv},d}(k) + \tfrac{T_s}{L_f}\left[v_{\mathrm{grid},q}(k) - v_{\mathrm{conv},q}(k)\right] \end{aligned}\right\} \tag{14}$$

The variable $v_{\mathrm{grid},dq}(k)$ can take eight different combinations [10] depending upon the switching states of three-phase two-level VSC, resulting in eight predicted values of $i_{\mathrm{grid},dq}(k + 1)$. MPC of GSC is shown in Fig. 3.

The cost function to be minimized is the error between the predicted current and the reference current expressed in *dq*-coordinated frame as under:

$$g = \left|i_{\mathrm{conv},d}^{\mathrm{ref}}(k) - i_{\mathrm{conv},d}^{\mathrm{pred}}(k+1)\right| + \left|i_{\mathrm{conv},q}^{\mathrm{ref}}(k) - i_{\mathrm{conv},q}^{\mathrm{pred}}(k+1)\right| \tag{15}$$

The constrained can be easily implemented in the cost function.

$$g = \left|i_{\mathrm{conv},d}^{\mathrm{ref}}(k) - i_{\mathrm{conv},d}^{\mathrm{pred}}(k+1)\right| + \left|i_{\mathrm{conv},q}^{ref}(k) - i_{\mathrm{conv},q}^{\mathrm{pred}}(k+1)\right| + f\left(i_{\mathrm{conv},d}^{\mathrm{pred}}(k+1), i_{\mathrm{conv},q}^{\mathrm{pred}}(k+1)\right) \tag{16}$$

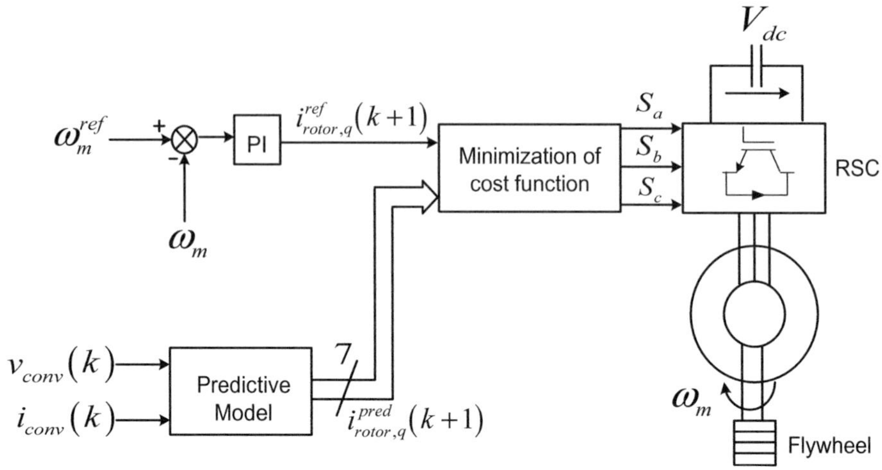

Fig. 4 MPC of rotor-side VSC

where the term $f\left(i_{\text{conv},d}^{\text{pred}}(k+1), i_{\text{conv},q}^{\text{pred}}(k+1)\right)$ is nonlinear in nature defined by:

$$f\left(i_{\text{conv},d}^{\text{pred}}(k+1), i_{\text{conv},q}^{\text{pred}}(k+1)\right) = \begin{cases} \infty, & \text{if } \left|i_{\text{conv},d}^{\text{pred}}(k+1)\right| > i_{\text{conv,max}} \text{ or } \left|i_{\text{conv},q}^{\text{pred}}(k+1)\right| > i_{\text{conv,max}} \\ 0, & \text{if } \left|i_{\text{conv},d}^{\text{pred}}(k+1)\right| \le i_{\text{conv,max}} \text{ and } \left|i_{\text{conv},q}^{\text{pred}}(k+1)\right| \le i_{\text{conv,max}} \end{cases} \quad (17)$$

The switching states of VSC is selected that leads to minimum value of cost function g.

4.2 *MPC of Rotor-Side VSC*

Predicted current at time instant $t = (k+1)$ for RSC is obtained by discretizing (11) and rearranging it as:

$$\left.\begin{aligned} i_{\text{rotor},d}(k+1) &= \left[1 - \tfrac{R_r T_s}{\sigma L_r}\right] i_{\text{rotor},d}(k) + \omega_{\text{slip}}(k) T_s i_{\text{rotor},q}(k) + \tfrac{T_s}{\sigma L_r} v_{\text{rotor},d}(k) \\ i_{\text{rotor},q}(k+1) &= \left[1 - \tfrac{R_r T_s}{\sigma L_r}\right] i_{\text{rotor},q}(k) - \omega_{\text{slip}}(k) \tfrac{T_s}{\sigma L_r} x \left[L_{mm} i_{ms}(k) + \sigma L_r i_{\text{rotor},d}(k)\right] + \tfrac{T_s}{\sigma L_r} v_{\text{rotor},q}(k) \end{aligned}\right\} \quad (18)$$

where $\omega_{\text{slip}} = (\omega_s - \omega_r)$ is the electrical slip speed of the rotor. MPC of RSC is shown in Fig. 4.

The cost function to be minimized is selected such that reactive power exchanged by DFIM with grid is zero, tracking the desired reference speed and limiting the amplitude of rotor current.

$$g = \left| i_{\text{rotor},d}^{\text{ref}}(k) - i_{\text{rotor},d}^{\text{pred}}(k+1) \right| + \left| i_{\text{rotor},q}^{\text{ref}}(k) - i_{\text{rotor},q}^{\text{pred}}(k+1) \right| + f\left(i_{\text{rotor},d}^{\text{pred}}(k+1), i_{\text{rotor},q}^{\text{pred}}(k+1) \right) \quad (19)$$

The nonlinear part of the cost function is defined in the same way as in (17).

4.3 Delay Compensation in MPC

Delay is introduced into the control process due to the prediction of currents of both GSC and RSC and in the cost function calculation since current prediction and cost calculation is to be done for all the voltage vectors of VSC which in the present case is eight in number. Moreover, the effect of sampling time and speed of microprocessor also contribute to the time delay. This delay is compensated through the method used in [11]. The following cost function is minimized for optimal voltage vectors.

$$g = \left| i_{\text{conv},d}^{\text{ref}}(k) - i_{\text{conv},d}^{\text{pred}}(k+2) \right| + \left| i_{\text{conv},q}^{\text{ref}}(k) - i_{\text{conv},q}^{\text{pred}}(k+2) \right| + f\left(i_{\text{conv},d}^{\text{pred}}(k+2), i_{\text{conv},q}^{\text{pred}}(k+2) \right) \quad (20)$$

$$g = \left| i_{\text{rotor},d}^{\text{ref}}(k) - i_{\text{rotor},d}^{\text{pred}}(k+2) \right| + \left| i_{\text{rotor},q}^{\text{ref}}(k) - i_{\text{rotor},q}^{\text{pred}}(k+2) \right| + f\left(i_{\text{rotor},d}^{\text{pred}}(k+2), i_{\text{rotor},q}^{\text{pred}}(k+2) \right) \quad (21)$$

where the terms $f\left(i_{\text{conv},d}^{\text{pred}}(k+2), i_{\text{conv},q}^{\text{pred}}(k+2) \right)$ and $f\left(i_{\text{rotor},d}^{\text{pred}}(k+2), i_{\text{rotor},q}^{\text{pred}}(k+2) \right)$ are defined in similar fashion as in (17).

5 Simulation Results

5.1 Speed Tracking Comparison

A combined inertia of flywheel and rotor of DFIM corresponding to 4.0 s is taken for study. The proportional constant $\left(K_p\right)$ and integral constant (K_i) for outer loop of PI–PI controlled FESS as well as for PI–MPC controlled FESS are kept equal. This corresponds to same bandwidth (BW) and damping ratio for outer loop and thus the overall time constant of PI–PI and PI–MPC controlled FESS are same. This control is then studied for different values of inner-loop time constant of PI–PI and is compared with PI–MPC.

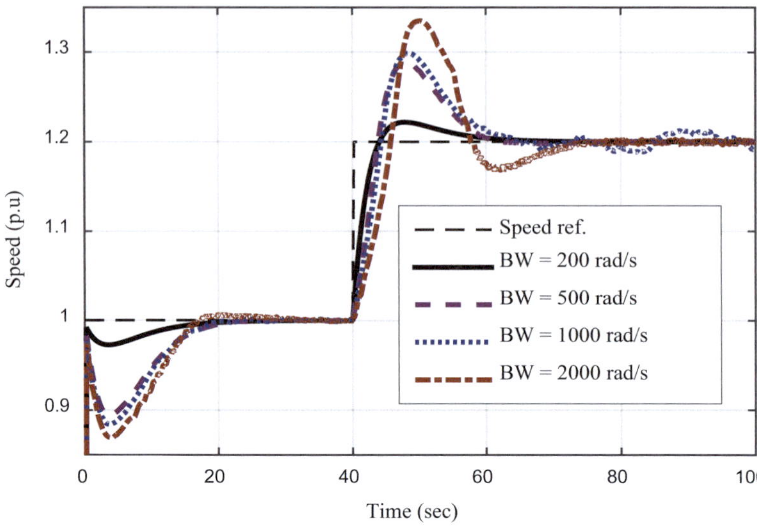

Fig. 5 Step change in speed from 1.0 to 1.2 pu

The time constant of the PI-based inner-loop control of RSC is varied from $\tau = 0.5$ ms to $\tau = 5.0$ ms [12] which corresponds to BW of 200 rad/s to 2000 rad/s, respectively. Figure 5 shows the comparison of speed output of flywheel for various BWs of inner-loop PI–PI controlled FESS. It is found that for 200 and 500 rad/s, the responses are the best. Hence, these are used in study.

A decrease in overshoot in speed is observed when the time constant of the inner loop is increased (BW is decreased) from $\tau = 0.5$ ms (BW = 2000 rad/s) to $\tau = 5.0$ ms (BW = 200 rad/s). This FESS is again simulated for various BWs of inner loop for a step fall in speed at $t = 40.0$ s. Figure 6 shows speed tracking of FESS for various values of inner-loop BW. It is again observed that as time constant is increased (BW is decreased) from $\tau = 0.5$ ms (BW = 2000 rad/s) to $\tau = 5.0$ ms (BW = 200 rad/s), overshoot in speed is reduced.

The inner loop of RSC of FESS is now controlled through MPC and speed tracking, and torque variation is now compared with PI-based control with time constant of inner loop as $\tau = 2.0$ ms (BW = 500 rad/s) and $\tau = 5.0$ ms (BW = 200 rad/s). Figure 7 shows the speed tracking performance of FESS for MPC and PI-based controller. It is seen that speed overshoot for MPC is least as compared to PI-based controller. Torque corresponding to this speed tracking is shown in Fig. 8. Maximum torque overshoot is equal to that PI-based controller for $\tau = 5.0$ ms.

For a step fall in speed from 1.0 to 0.8 pu at $t = 40$ s, speed tracking performance of MPC is again compared with PI as is shown in Fig. 9. It is again observed that MPC-based torque control gives less speed overshoot than PI-based control (Fig. 10).

In PI-based control, as the BW of the overall controlled process is increased, the settling time is reduced; however, the speed overshoot is increased. This is due to the increase in integral constant of controller. As MPC is a high proportional gain control

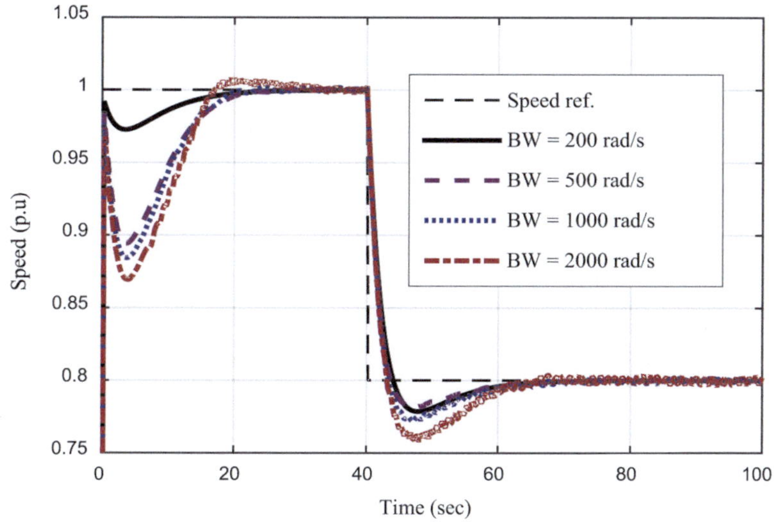

Fig. 6 Step change in speed from 1.0 to 0.8 pu

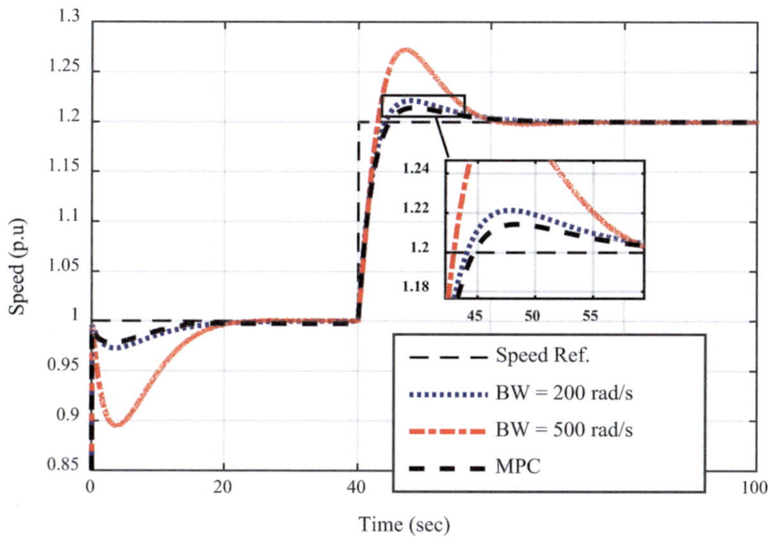

Fig. 7 Speed tracking performance under step rise in speed between MPC and PI controller

with no integral action, speed overshoot is least during speed tracking as shown in Figs. 7 and 9. A small amount of speed overshoot in MPC-based control is due to outer PI-loop control of speed.

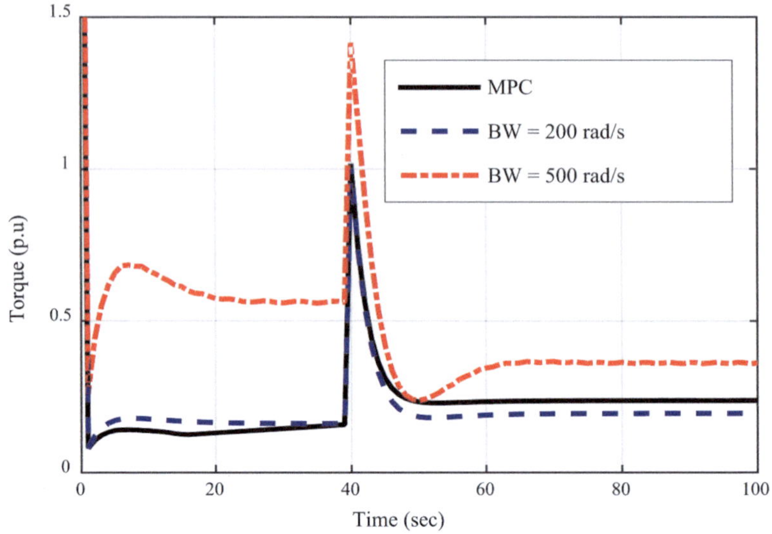

Fig. 8 Torque corresponding to speed tracking

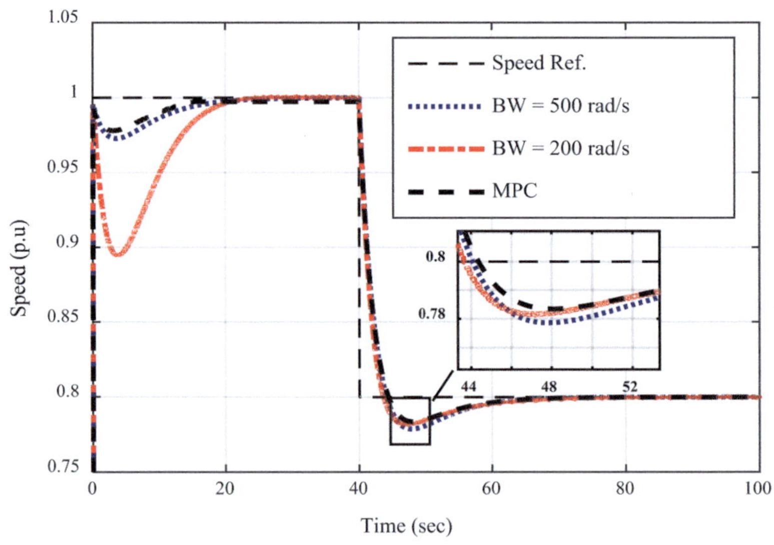

Fig. 9 Speed tracking performance under step fall in speed between MPC and PI controller

5.2 Active Power Regulation Using FESS

A stochastic wind speed is simulated using random value function where wind speed is varied from 7 to 14 m/s with an average value of 10.5 m/s as shown in Fig. 11.

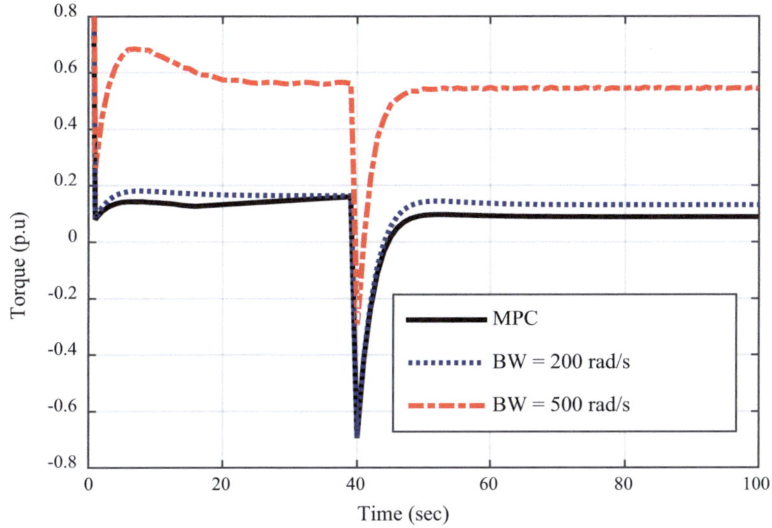

Fig. 10 Torque corresponding to speed tracking

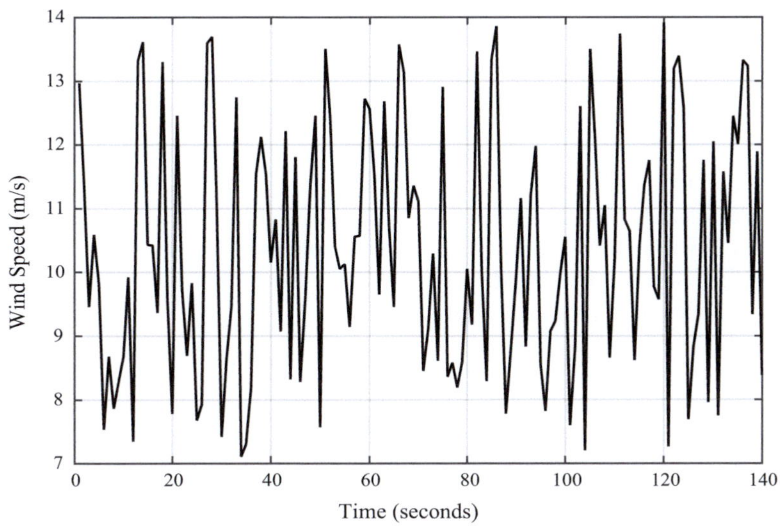

Fig. 11 Wind speed

The power output from the wind turbine without FESS and with FESS is simulated. It is seen from Fig. 12 that with FESS, the power generated by wind turbine is better regulated.

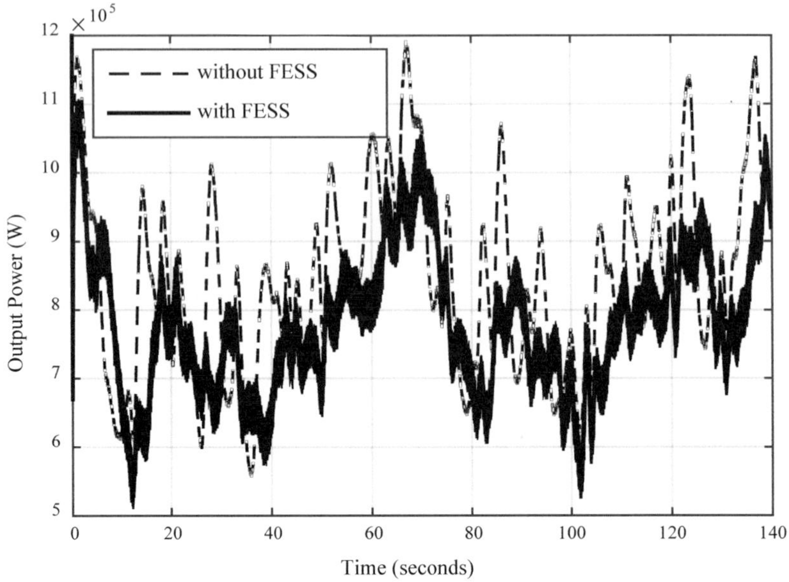

Fig. 12 Wind power output with and without FESS

6 Conclusion

In this paper, an integrated operation of FESS based on MPC control with DFIG is studied for active power regulation through FESS. During varying wind conditions, the output power of DFIG is varying. This varying output power is then regulated through FESS. Speed tracking through MPC is better in both step rise in speed and step fall in speed as compared to PI-based control.

References

1. Barton, J.P., Infield, D.G.: Energy storage and its use with intermittent renewable energy. IEEE Trans Energy Convers. **19**(2):441–448 (2004)
2. Faraji, F., Majazi, A., Al-Haddad, K.: A comprehensive review of flywheel energy storage system technology. Renew. Sustain. Energy Rev. **67**, 477–490 (2017)
3. Suvire, G.O., Mercado, P.E.: Active power control of a flywheel energy storage system for wind energy applications. IET Renew. Power Gener. **6**(1), 9–16 (2012)
4. Cimuca, G., Breban, S., Radulescu, M.M., Saudemont, C., Robyns, B.: Design and control strategies of an induction-machine-based flywheel energy storage system associated to a variable-speed wind generator. IEEE Trans. Energy Convers. **25**(2), 526–534 (2010)
5. Ghosh, S., Kamalasadan, S.: An energy function-based optimal control strategy for output stabilization of integrated DFIG-Flywheel energy storage system. IEEE Trans. Smart Grid (2016)

6. Daoud, M.I., Massoud, A.M., Abdel-Khalik, A.S., Elserougi, A., Ahmed, S.: A flywheel energy storage system for fault ride through support of grid-connected VSC HVDC-based offshore wind farms. IEEE Trans. Power Syst. **31**(3), 1671–1680 (2016)
7. Diaz-Gonzalez, F., Bianchi, F.D., Sumper, A., Gomis-Bellmunt, O.: Control of a flywheel energy storage system for power smoothing in wind power plants. IEEE Trans. Energy Convers. **29**(1), 204–214 (2014)
8. Abad, G., Lopez, J., Rodríguez, M., Marroyo, L., Iwanski, G.: Doubly Fed Induction Machine: Modeling and Control for Wind Energy Generation, vol. 85. Wiley, New York (2011)
9. Camacho, E.F., Bordons, C.: Model Predictive Control in the Process Industry. Springer Science & Business Media (2012)
10. Rodriguez, J., Cortes, P.: Predictive Control of Power Converters and Electrical Drives, vol. 40. Wiley, New York (2012)
11. Cortes, P., Rodriguez, J., Silva, C., Flores, A.: Delay compensation in model predictive current control of a three-phase inverter. IEEE Trans. Industr. Electron. **59**(2), 1323–1325 (2012)
12. Yazdani, A., Iravani, R.: Voltage-Sourced Converters in Power Systems: Modeling, Control, and Applications. Wiley, New York (2010)

Post-evaluation Index System and Comprehensive Evaluation Method for Daily Power System Dispatching

Pengcheng Cui, Bing Xu, Fushuan Wen, Zihao Fu, Bingquan Zhu, Qiulong Ni and S. N. Singh

Abstract Since the operating characteristics of a modern power system is becoming increasingly complicated, it is imperative to assess the daily power system dispatching quality. First, based on the average effect, buckets effect and abnormalities effect, a post-evaluation index system for daily power system dispatching is presented from the perspectives of system security, economics, energy saving, environmental friendliness, and impartiality. Meanwhile, a comprehensive evaluation method is proposed to integrate the evaluation indexes. The scores of the indexes are calculated by the well-established extreme value method, and the index weights are divided into subjective and objective ones which are determined by the G1 method, an extended version of the well-known analytic hierarchy process (AHP) method, combined with the fuzzy clustering method and improved entropy method, respectively. Following the minimum identification information principle, the comprehensive weights can

P. Cui · F. Wen (✉)
College of Electrical Engineering, Zhejiang University, Hangzhou 310027, China
e-mail: fushuan.wen@gmail.com

P. Cui
e-mail: daniel.cui.zju@qq.com

B. Xu
State Grid Zhejiang Electric Power Corporation Economic Research Institute, Hangzhou 310008, China
e-mail: xubingzju@hotmail.com

Z. Fu
China Development Bank Ningbo Branch, Ningbo 315040, China
e-mail: fzh5224@sina.com

B. Zhu · Q. Ni
State Grid Zhejiang Electric Power Corporation, Hangzhou 310007, China
e-mail: hzzbq@sina.com

Q. Ni
e-mail: 10131415@qq.com

S. N. Singh
Department of Electrical Engineering, Indian Institute of Technology Kanpur, Kanpur 208016, India
e-mail: snsingh@iitk.ac.in

S. N. Singh et al. (eds.), *Advances in Energy and Power Systems*, Lecture Notes in Electrical Engineering 508, https://doi.org/10.1007/978-981-13-0662-4_7

be determined, and hence, the final assessment results attained by the weighted average value of index scores. Case studies show that the proposed method cannot only identify weak links, but also reflect the overall level accurately, and it is feasible and effective for applications in practical power systems.

Keywords Daily power system dispatching · Evaluation index system · G1-fuzzy clustering method · Improved entropy method

1 Introduction

In recent years, due to the downward pressure of economics in China, the growth rate of the overall electrical demand has slowed down. Electric power companies will progressively enter a new stage of lean development. In addition, with the evergrowing scale of a modern power system, its operating characteristic is becoming increasingly complex. Hence, there is a need to establish a post-evaluation index system so as to identify the weak links in the power system and improve the lean level of the daily power system dispatching process.

Up to now, much research work has been done on the evaluations of many aspects in power systems. In Billinton and Wangdee [1], an index system for evaluating power system reliability is proposed by utilizing sequential Monte Carlo simulations. A real-time assessment tool for power system security is proposed by Diao et al. [2], based on phasor measurement units (PMUs) and decision trees (DTs). Salarvand et al. [3] employed the well-known artificial neural network (ANN) and fuzzy logic to obtain global indices for power quality evaluation. A hierarchical and partition low-carbon index system is established by Kerin et al. [4] and can perform assessments on power generation, transmission and distribution, and customer sectors. In Wang et al. [5], an index system is presented with reliability, economics, power quality, energy efficiency, and fairness included, but the evaluation method is not involved. The assessments on several aspects of smart grids have been carried out in recent years such as the development level and possible impacts [6–8]; the level and scale of new technology as well as benefits brought by smart grid development. To the best of our knowledge, the post-evaluation for daily power system dispatching has not yet been systematically examined in existing publications.

Given the aforementioned background, based on the data provided by a power system dispatching support system, a post-evaluation index system for daily power system dispatching is proposed from the perspectives of power system security, economics, energy saving, environmental friendliness, and impartiality. The subjective weights and the objective weights are obtained by the G1 method combined with the fuzzy clustering (FC) method and the improved entropy method, respectively. Following the minimum identification information principle, the comprehensive weights are determined for aggregating the established indexes. The evaluation method not only excavates the subjective experience of domain experts, but also considers the information content of the available data.

The remainder of this paper is organized as follows. A brief introduction of the principles in establishing an index system is given in Sect. 2. In Sect. 3, the post-evaluation index system for daily power system dispatching is addressed. In Sect. 4, an evaluation model based on the combination weighting method is proposed. Case studies are given in Sect. 5 to demonstrate the proposed index system and evaluation method, and the paper is concluded in Sect. 6.

2 Principles of Establishing a Post-evaluation Index System

Unlike the evaluation of a single device or the real-time operating status of a given power system, the post-evaluation for daily power system dispatching needs to coordinate the contradiction between the individuals and the whole, as well as the contradiction between real-time data and statistical data in a given period. Therefore, the following effects should also be taken into consideration, in addition to the systematic, scientific, measurable, and comparable principles:

Average effect. When the status of objects is in their normal range, the average value is utilized to reflect the overall level of the objects.

Buckets effect. When there is an abnormal individual in the objects, the evaluation index mainly depends on this abnormal individual.

Abnormalities effect. When there are multiple abnormal individuals, in addition to the worst individual, other abnormal individuals need to be taken into account as they have an impact on the evaluation index also.

3 Post-evaluation Index System for Daily Power System Dispatching

Based on the stated principles in the previous section, a post-evaluation index system for daily power system dispatching is established. The index system consists of five first-class indexes on power system security, economics, energy saving, environmental friendliness, and impartiality as shown in Table 1.

3.1 Security

The short-circuit current level ($x_{1.3}$). To judge whether the short circuit is beyond the given limit, $x_{1.3}$ is defined as

Table 1 Post-evaluation index system for daily power system dispatching

First-class index	Second-class index
Security, x_1	N-1 passing rate, $x_{1.1}$
	N-2 passing rate, $x_{1.2}$
	Short-circuit current level, $x_{1.3}$
	The comprehensive load ratio of transmission sections, $x_{1.4}$
	The comprehensive load ratio of main transformers, $x_{1.5}$
	The comprehensive load ratio of transmission lines, $x_{1.6}$
	Power flow nonuniformity degree, $x_{1.7}$
	The rate of insufficient spinning reserve, $x_{1.8}$
	Reactive power margin, $x_{1.9}$
	The accuracy of load forecasting, $x_{1.10}$
	Voltage qualification rate, $x_{1.11}$
	Frequency qualification rate, $x_{1.12}$
Economics, x_2	Average power purchase cost deviation with respect to perfect dispatch, $x_{2.1}$
	Line loss rate, $x_{2.2}$
	The rate of excessive spinning reserve, $x_{2.3}$
	Unit load ratio under peak load period, $x_{2.4}$
	Reactive power balance degree, $x_{2.5}$
Energy saving, x_3	The average coal consumption for generating per unit electricity, $x_{3.1}$
	Average load factor of generating units, $x_{3.2}$
Environmental friendliness, x_4	Grid-connected rate of renewable energy units, $x_{4.1}$
	Generation share by renewable energy, $x_{4.2}$
	The emitted SO_2 quantity for generating per MW electricity, $x_{4.3}$
	The emitted N_xO_x quantity for generating per MW electricity, $x_{4.4}$
	The emitted fly ash quantity for generating per MW electricity, $x_{4.5}$
Impartiality, x_5	The deviation of daily generation scheduling completion rate, $x_{5.1}$
	The nonuniformity degree of load factors among generating units, $x_{5.2}$

$$x_{1.3} = \begin{cases} \max\left(\frac{I_i}{I_{i,\max}}\right) + \frac{1}{n_e}\sum_{i=1}^{n_{ey}} \frac{I_i}{I_{i,\max}} & \exists \frac{I_i}{I_{i,\max}} \geq 1 \\ \frac{1}{n_e}\sum_{i=1}^{n_e} \frac{I_i}{I_{i,\max}} & \forall \frac{I_i}{I_{i,\max}} < 1 \end{cases} \quad (1)$$

where I_i is the larger short-circuit current between the single-phase grounding fault and the three-phase short circuit fault on bus i; $I_{i,\max}$ is the rated short-circuit breaking current of the circuit breaker on bus i; n_e and n_{ey} are respectively the numbers of buses and the over-current limit buses.

The Comprehensive Load Ratio of Transmission Sections ($x_{1.4}$). The overload of transmission sections has a strong correlation with the occurrence of chain failure. Therefore, it is necessary to monitor the power flow of transmission sections.

$$x_{1.4} = \begin{cases} \max\limits_{288\text{ points}}\left[\max\left(\frac{P_{Si}}{P_{Si,\max}}\right) + \frac{1}{n_S}\sum_{i=1}^{n_{sy}} \frac{P_{Si}}{P_{Si,\max}}\right] & \exists \frac{P_{Si}}{P_{Si,\max}} \geq 1 \\ \max\limits_{288\text{ points}}\left(\frac{1}{n_s}\sum_{i=1}^{n_s} \frac{P_{Si}}{P_{Si,\max}}\right) & \forall \frac{P_{Si}}{P_{Si,\max}} < 1 \end{cases} \quad (2)$$

where the number 288 refers to the sampling points in a day with 5-min sampling durations; P_{si} is the current actual active power flow of transmission sections i; $P_{si,\max}$ is the capacity limit of transmission section i; n_s and n_{sy} are respectively the numbers of transmission sections and the over-current limit ones.

Power Flow Nonuniformity Degree ($x_{1.7}$). This index reflects the nonuniformity degree of power flow among all branches. At the same load level, the smaller the index value is, the better the uniformity degree of power flow among all branches is.

$$x_{1.7} = \max_{288\text{ points}}\left[\frac{1}{n_l}\sum_{i=1}^{n_l}\left(R_{li} - R_{l,\text{ave}}\right)^2\right] \quad (3)$$

$$R_{li} = I_{li}/I_{li,\max} \quad (4)$$

where R_{li} and $R_{l,\text{ave}}$ are respectively the load ratio of branch i and the present average load ratio among all branches; I_{li} and $I_{li,\max}$ are the current value and current limit of branch i, respectively; n_l is the number of branches.

Reactive Power Margin ($x_{1.9}$). In a given region, the available reactive power capacity should be enough for supplying reactive power load. It is necessary to maintain a certain degree of reserve so as to meet the increased requirement of reactive load and maintain acceptable voltage profile. The reactive power margin index can be expressed as

$$x_{1.9} = \begin{cases} \min\limits_{288\text{ points}}\left[\min\left(\frac{Q_{\mathrm{G}i}-Q_{\mathrm{L}i}}{Q_{\mathrm{G}i}}\right)\right] & \forall \frac{Q_{\mathrm{L}i}}{Q_{\mathrm{G}i}} < 1 \\ 0 & \exists \frac{Q_{\mathrm{L}i}}{Q_{\mathrm{G}i}} > 1 \end{cases} \quad (5)$$

where $Q_{\mathrm{L}i}$ is the current total reactive load of region i; $Q_{\mathrm{G}i}$ is the available total reactive power capacity in region i.

3.2 Economics

Average Power Purchase Cost Deviation with Respect to Perfect Dispatch ($x_{2.1}$). In order to accurately reflect the degree of the average purchase cost that can be reduced by dispatching, this paper adopts the concept of perfect dispatch (PD) proposed by the Pennsylvania–New Jersey–Maryland (PJM) [9], which refers to the least production cost under the premise that all system conditions are known and controllable. The average power purchase cost deviation is defined as

$$x_{2.1} = \frac{C_{\mathrm{rea}} - C_{\mathrm{per}}}{C_{\mathrm{per}}} \times 100\% \quad (6)$$

where C_{rea} and C_{per} are respectively the average power purchase cost under the actual dispatching and PD.

Reactive Power Balance Degree ($x_{2.5}$). If the reactive power transmits over a long distance or through transformers, it will cause significant loss; hence, it should be balanced hierarchically and divisionally. Taking power factor of the main transformers' high voltage side in 220 kV substations as assessment unit, the reactive power balance degree is defined as:

$$x_{2.5} = \min_{j \in n_{\mathrm{tc}}}\left(\frac{1}{n_{\mathrm{T}j}} \sum_{i=1}^{n_{\mathrm{T}j}} \frac{N_{\phi i}}{288}\right) \quad (7)$$

where $N_{\phi i}$ is the number of sampling points of qualified power factor for substation i; n_{tc} and $n_{\mathrm{T}j}$ are respectively the number of regions and substations in region j.

3.3 Energy Saving

Average Load Factor of Generating Units ($x_{3.2}$). When the generator unit is running at the high load factor, the coal consumption is low. Hence, it is significant to improve the load factor. To evaluate the load factor of generating units, the index is defined as:

$$x_{3.2} = \frac{\sum_{i=1}^{n_{\mathrm{G}}} a_{\mathrm{G}i} L_i}{n_{\mathrm{G}}}$$

$$\alpha_{\mathrm{G}i} = \begin{cases} 1.2, C_i > 600\ \mathrm{MW} \\ 1.1, 600\ \mathrm{MW} \geq C_i > 450\ \mathrm{MW} \\ 0.9, 450\ \mathrm{MW} \geq C_i > 300\ \mathrm{MW} \\ 0.8, C_i \leq 300\ \mathrm{MW} \end{cases} \tag{8}$$

where n_{G} is the number of units; L_i and C_i are respectively the daily load factor and the capacity of unit i; $a_{\mathrm{G}i}$ is the weight of unit i, which is set to encourage high capacity units and its value refers to (8).

3.4 *Impartiality*

The Deviation of Daily Generation Scheduling Completion Rate ($x_{5.1}$). The evaluation index refers to deviation of the ratio of the real electricity generation to the planned and is defined as:

$$x_{5.1} = \frac{1}{n_{\mathrm{G}}} \sum_{i=1}^{n_{\mathrm{G}}} (D_i - D_{\mathrm{ave}})^2 \tag{9}$$

where D_i and D_{ave} denote the completion rate of daily generation scheduling of the unit i and the average value, respectively.

The Nonuniformity Degree of Load Factors among Generating Units ($x_{5.2}$).

$$x_{5.2} = \frac{\sum_{i=1}^{n_{\mathrm{G}}} |\beta_{\mathrm{G}i} L_i - L_{\mathrm{ave}}|}{n_{\mathrm{G}}}$$

$$\beta_{\mathrm{G}i} = \begin{cases} 0.8, C_i > 600\ \mathrm{MW} \\ 0.9, 600\ \mathrm{MW} \geq C_i > 450\ \mathrm{MW} \\ 1.1, 450\ \mathrm{MW} \geq C_i > 300\ \mathrm{MW} \\ 1.2, C_i \leq 300\ \mathrm{MW} \end{cases} \tag{10}$$

where L_{ave} represents the average load factor of units; $\beta_{\mathrm{G}i}$ is the weight to ensure that the load factor of high capacity units is higher than the low.

4 Comprehensive Evaluation Method Based on Combination Weighting Method

Each affiliated index needs to be integrated to evaluate the overall level of daily power system dispatching. The indexes are first transformed into unified index scores, and then the combination weighting method is utilized to determine the optimal weight of each index. The evaluation result is obtained by weighted aggregation of index scores finally.

4.1 Index Score Based on Extreme Value Method

According to the relationship between the index value and the expected value, the indexes can be divided into cost-type indexes and benefit-type indexes. The smaller the value of the cost-type index, the greater the score, and benefit-type index otherwise. Extreme value method is adopted to normalize the original data.

(1) Score function of cost-type indexes

$$r_{ij} = \frac{x_{i,\max} - x_{ij}}{x_{i,\max} - x_{i,\min}} \times 100 \tag{11}$$

where r_{ij} and x_{ij} are respectively the score and the value of the evaluation index i on day j; $x_{i,\max}$ and $x_{i,\min}$ are respectively the historical maximum and minimum values of the evaluation index i.

(2) Score function of benefit-type indexes

$$r_{ij} = \frac{x_{ij} - x_{i,\min}}{x_{i,\max} - x_{i,\min}} \times 100 \tag{12}$$

4.2 Set of the Objective Weight

The G1 method is a weighting method developed from the analytic hierarchy process (AHP). Compared with AHP, it does not need to construct the judgment matrix or check the consistency; hence, the computation burden can be mitigated [10]. Owing to space constraints, specific calculation steps of the G1 method are not given here and are available in Yun-Nai et al. [10]. However, the most significant elements as well as the order may vary, because each expert refers to his/her personal experience, knowledge, views, and other factors of assigning significances to the evaluation indexes. Hence, the fuzzy clustering (FC) method is employed to cluster the experts' opinion [11]. The calculation process of the G1-FC weighting method is detailed below:

1) n experts are employed to determine the weights of m evaluation indexes using the G1 method, and the weight matrix is shown as

$$\mathbf{W} = \begin{bmatrix} w_{1,1} & w_{1,2} & \cdots & w_{1,n} \\ w_{2,1} & w_{2,2} & \cdots & w_{2,n} \\ \vdots & \vdots & \cdots & \vdots \\ w_{m,1} & w_{m,2} & \cdots & w_{m,n} \end{bmatrix} \tag{13}$$

where $w_{i,j}$ is the weight given by expert j for evaluation index i.

(2) The opinion similarity $d(x,y)$ between expert x and expert y is determined by the angle cosine as

$$d(x, y) = \frac{\sum_{i=1}^{m} w_{i,x} w_{i,y}}{\sqrt{\left(\sum_{i=1}^{m} w_{i,x}^2\right) \times \left(\sum_{i=1}^{m} w_{i,y}^2\right)}} \tag{14}$$

The similarity matrix $\mathbf{D} = \left[d(x, y)_{n \times n}\right]$ can be determined according to the pairwise comparison of n experts.

(3) If the similarity of two experts is larger than the threshold value T, then they are allocated to the same cluster. Assume that experts are classified into l clusters and the number of experts in cluster k is Φ_k.
(4) The subjective weight given by expert j in cluster k is $\mathbf{W}_{jk} = [w_{1,jk}\ w_{2,jk}, \ldots\ w_{m,jk}]^{\mathrm{T}}$. For a large-capacity cluster, a greater inter-cluster weight should be set as it represents the opinion of the majority. For the intra-cluster weight, the smaller the information entropy of the weight vector is, the more definite idea this expert has [12]. Therefore, a larger intra-cluster weight should be set.

$$H(\mathbf{W}_{jk}) = -\frac{1}{\ln m} \sum_{i=1}^{m} w_{i,jk} \ln\left(w_{i,jk}\right) \tag{15}$$

$$\lambda_k = \Phi_k^2 / \sum_{g=1}^{l} \Phi_g^2 \tag{16}$$

$$\alpha_{jk} = \left[1 - H(\mathbf{W}_{jk})\right] / \sum_{g=1}^{\Phi_k} \left[1 - H(\mathbf{W}_{gk})\right] \tag{17}$$

where $H(\mathbf{W}_{jk})$ is the information entropy of expert j in cluster k; λ_k is the inter-cluster weight of cluster k; α_{jk} is the intra-cluster weight of expert j in cluster k.

(5) The subjective weight $w_{s,i}$ can be obtained as

$$w_{s,i} = \sum_{k=1}^{l} \sum_{j=1}^{\Phi_k} \lambda_k \alpha_{jk} w_{i,jk} \tag{18}$$

4.3 Set of Subjective Weights

The entropy method determines index weights according to the amount of effective information contained in each index. However, in the traditional entropy weight method, the weight is sensitive to information entropy. When the information entropy approximates 1, a small difference can cause the change of the index weight doubly which lead to the inconsistent situation between index importance and weight [13]. Therefore, this paper utilizes the improved entropy weight method to overcome this shortcoming. There are m evaluation indexes and q evaluation days. The specific steps are as follows:

(1) The entropy value H_i of the index i is:

$$H_i = -\frac{1}{\ln q} \sum_{j=1}^{q} p_{ij} \ln p_{ij} \tag{19}$$

where $p_{ij} = r_{ij} / \sum_{j=1}^{q} r_{ij}$; $p_{ij} \ln p_{ij} = 0$ when $p_{ij} = 0$.

(2) The objective weight $w_{o,i}$ of the index i is:

$$w_{o,i} = \begin{cases} \left(1 - \overline{H}^{35.35}\right) w_{0i} + \overline{H}^{35.35} w_{3i} & H_i < 1 \\ 0 & H_i = 1 \end{cases} \tag{20}$$

$$\begin{cases} w_{0i} = \frac{1-H_i}{\sum_{i=1}^{q} (1-H_i)} \\ w_{3i} = \frac{1+\overline{H}-H_i}{\sum_{k=1, H_k \neq 1}^{q} (1+\overline{H}-H_k)} \end{cases} \tag{21}$$

where $\overline{H}$ is the average of all entropy values that are not 1.

4.4 Calculation of Comprehensive Weights

To make full use of the experts' knowledge and the sample data information, the principle of minimum discriminant information is adopted to determine the comprehensive weight $\boldsymbol{w}_c$. To make sure that $\boldsymbol{w}_c$ is as close to the subjective weight $\boldsymbol{w}_s$ and the objective weight $\boldsymbol{w}_o$ as possible, the objective function and constraint condition can be formulated as

$$\min F = \sum_{i=1}^{m} w_{c,i}\left(\ln w_{c,i} - \ln w_{s,i}\right) + \sum_{i=1}^{m} w_{c,i}\left(\ln w_{c,i} - \ln w_{o,i}\right) \tag{22}$$

$$\text{s.t.} \sum_{i=1}^{m} w_{c,i} = 1; \; w_{c,i} > 0, \quad i = 1, 2, \ldots, m \tag{23}$$

The Lagrange multiplier method is used to solve the above optimization problem, and $w_{c,i}$ is expressed as

$$w_{c,i} = \frac{\left[w_{s,i} w_{o,i}\right]^{0.5}}{\sum_{i=1}^{m} \left[w_{s,i} w_{o,i}\right]^{0.5}} \quad i = 1, 2, \ldots, m \tag{24}$$

The comprehensive evaluation result of the day j can be obtained according to the index scores and the comprehensive weights.

$$G_j = \sum_{i=1}^{m} r_{ij} w_{c,i} \tag{25}$$

where G_j is the evaluation result of the day j.

5 Case Study

A provincial power system is taken as an example. Due to the limited space, only part of the calculation process is listed.

5.1 Comprehensive Weight of Economics Indexes

The economics indexes are employed to illustrate the detailed calculation process of index weights.

Calculation of index scores. The basic data can be obtained from the power system dispatching control system, and index scores are calculated by the extreme value method. The average power purchase cost deviation with respect to perfect dispatch, the line loss rate, and the rate of excessive spinning reserve is cost-type indexes. The unit load rate of under peak load period and the reactive power balance degree is benefit-type indexes.

Table 2 Weight values of economics indexes

Weight type	$x_{2.1}$	$x_{2.2}$	$x_{2.3}$	$x_{2.4}$	$x_{2.5}$
Subjective weight	0.33	0.28	0.12	0.11	0.16
Objective weight	0.35	0.14	0.20	0.26	0.05
Comprehensive weight	0.36	0.21	0.16	0.18	0.09

Determination of subjective weights. The G1 method is adopted by experts separately to determine the subjective weights of the economics second-class indexes. The weight matrix **W** is

$$\mathbf{W} = \begin{bmatrix} 0.39\ 0.27\ 0.38\ 0.36\ 0.24\ 0.38\ 0.24 \\ 0.24\ 0.38\ 0.17\ 0.25\ 0.38\ 0.23\ 0.38 \\ 0.12\ 0.09\ 0.09\ 0.11\ 0.17\ 0.14\ 0.12 \\ 0.09\ 0.11\ 0.12\ 0.10\ 0.12\ 0.13\ 0.10 \\ 0.16\ 0.15\ 0.24\ 0.18\ 0.10\ 0.12\ 0.17 \end{bmatrix}$$

The similarity matrix between experts is calculated according to (14). The threshold value T is assumed to be 0.99. The clusters obtained are $\{(1,4,6)\ (2,7)\ (3)\ (5)\}$, and the inter-cluster weight and intra-cluster are as follows:

$$\lambda = [9/15\ 4/15\ 1/15\ 1/15]$$
$$\alpha_{11} = 0.38\ \alpha_{21} = 0.32\ \alpha_{31} = 0.30$$
$$\alpha_{12} = 0.55\ \alpha_{22} = 0.45\ \alpha_{13} = \alpha_{14} = 1$$

Hence, by employing (18) the subjective weight vector $\mathbf{w}_s$ of indexes is attained as [0.33, 0.28, 0.12, 0.11, 0.16].

Determination of objective weights. Five typical days are selected as samples. The score matrix **R** is as follows:

$$\mathbf{R} = \begin{bmatrix} 35\ 87\ 62\ 12\ 72 \\ 46\ 64\ 53\ 32\ 84 \\ 75\ 36\ 42\ 51\ 20 \\ 91\ 49\ 59\ 42\ 19 \\ 63\ 78\ 67\ 82\ 85 \end{bmatrix}$$

The objective weight of each index shown in Table 2 can be obtained by (19)–(21). After obtaining the subjective weight and the objective weight, the comprehensive weight of each index shown in Table 2 can be obtained by (25). The comprehensive weight combines the experience of experts and information of data simultaneously.

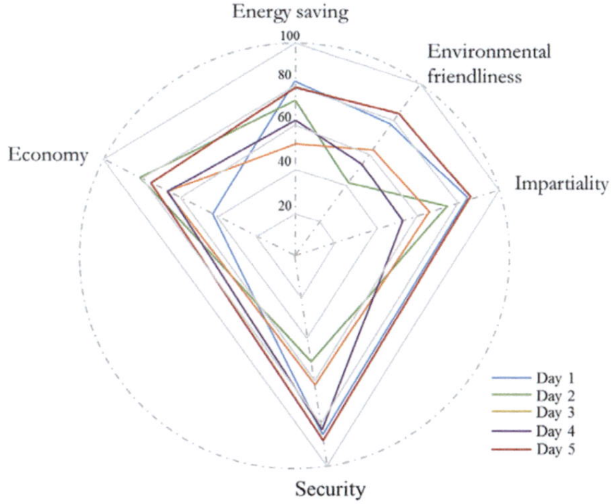

Fig. 1 Weighted radar chart for the assessment results of typical days

5.2 Evaluation Results of Typical Days

The evaluation result is illustrated in Fig. 1 utilizing weighted radar chart to reflect the weight of each index and the impact between each other [14].

The high load of Day 2 results in the overload of some transmission sections, main transformers, and lines. Because of the buckets effect and the abnormalities effect, the security score of Day 2 falls significantly. The economics of Day 1 is poor shown in Fig. 1. The main reason is that the rate of excessive spinning reserve is high, and the unit load ratio under peak load period is low. Therefore, in the subsequent dispatching process, it is recommended to reduce the spinning reserve capacity and adjust the unit startup and shutdown schedule. The energy saving index change along with the daily power system dispatching schedule. Environmental friendliness index has a certain stochastic fluctuation and weak correlation with the other indexes. Renewable energy is supposed to be accommodated preferentially to improve the environmental friendliness of grid operation. Impartiality index fluctuates by the daily dispatch schedule and the actual completion rate. For the situation of Day 4, it is necessary to allocate the generation scheduling for each unit reasonably.

The dispatchers can analyze the weak links in the dispatching and operation of the previous day according to the evaluation result and take corresponding measures to improve. The evaluation result of Day 5 demonstrates the validity.

6 Conclusion

To meet the requirement of lean dispatching and operation, a post-evaluation index system regarding security, economics, energy saving, environmental friendliness, and impartiality is first presented for daily power system dispatching considering the average effect, buckets effect, and abnormalities effect. A comprehensive evaluation method for evaluation index integration is next proposed; the extreme value method is adopted to determine the index score, while weights are properly determined by the G1 method combined with the fuzzy clustering method or improved entropy method. Finally, the evaluation result is attained by taking the weighted average value.

The daily power system dispatching quality of the previous day can be well reflected through the post-evaluation index system and comprehensive evaluation method proposed in this paper.

Acknowledgements This work is jointly supported by National Natural Science Foundation of China (51477151), National Basic Research Program of China (973 Program) (No. 2013CB228202), and a key project from State Grid Corporation of China (52110415001G).

References

1. Billinton, R., Wangdee, W.: Predicting bulk electricity system reliability performance indices using sequential Monte Carlo simulation. IEEE Trans. Power Deliv. **21**(2), 909–917 (2006)
2. Diao, R., Vittal, V., Logic, N.: Design of a real-time security assessment tool for situational awareness enhancement in modern power systems. IEEE Trans. Power Syst. **25**(2), 957–965 (2010)
3. Salarvand, A., Mirzaeian, B., Moallem, M.: Obtaining a quantitative index for power quality evaluation in competitive electricity market. IET Gen. Trans. Distrib. **4**(7), 810–823 (2010)
4. Kerin, U., Heyde, C., Krebs, R., Lerch, E.: Real-time dynamic security assessment of power grids. Eur. Phys. J. Spec. Top. **223**(12), 2503–2516 (2014)
5. Wang, H., Lin, Z., Wen, F., Huang, J.: A Comprehensive evaluation index system for power system operation. In: International Conference on Sustainable Power Generation and Supply (SUPERGEN 2012), pp. 1–6. IET Press, Hangzhou (2012)
6. Xenias, D., Axon, C.J., Whitmarsh, L., Connor, P.M., Balta-Ozkan, N., Spence, A.: UK smart grid development: an expert assessment of the benefits, Pitfalls and Functions. Renew. Energy **81**, 89–102 (2015)
7. Colak, I., Kabalci, E., Fulli, G., Lazarou, S.: A survey on the contributions of power electronics to smart grid systems. Renew. Sustain. Energy Rev. **47**, 562–579 (2015)
8. Livieratos, S., Vogiatzaki, V.E., Cottis, P.G.: A generic framework for the evaluation of the benefits expected from the smart grid. Energies **6**(2), 988–1008 (2013)
9. Chen, H.F., Bresler, S.: Practices on real-time market operation evaluation. IET Gen. Trans. Distrib. **4**(2), 324–332 (2010)
10. Yun-Nai, W., Yao, N., Huang, Z.: Application of fuzzy synthetic evaluation model based on entropy and G1 method in project management contract evaluation. In: International Conference on Management Science and Engineering, 2007 (ICMSE 2007), pp. 2284–2292. IEEE Press, Harbin (2007)
11. Oliveira, J.V.D., Pedrycz, W.: Advances in Fuzzy Clustering and Its Applications. Wiley, New York (2007)
12. Shannon, C.E.: A mathematical theory of communication. Bell Syst. Tech. J. **27**(3), 379–423 (1948)
13. Lu, Q., Zhao, Q.: Data processing in project supply chain vendor selection based on improved entropy weight—grey correlation—TOPSIS. Adv. Mater. Res. **1046**, 545–549 (2014)
14. Jeong, S., Jo, Y.M., Shim, S.O., Choi, Y.J., Youn, C.H.: A novel model for metabolic syndrome risk quantification based on areal similarity degree. IEEE Trans. Bio-Med. Eng. **61**(3), 665–679 (2014)

Real Assessment of State Owned Indian Electric Utilities

R. V. S. Sengar, Kalyan Chattterjee and Jay Singh

Abstract This paper presents a technique for calculating comparative cost efficiencies of Indian State Owned Electric Utilities (SOEUs), which have been primarily accountable for the distribution of electricity in India. Performance of twenty-eight state utilities was evaluated using the nonparametric approach of Data Envelopment Analysis (DEA). In this paper, input-oriented DEA is applied to calculate the overall technical and scale efficiencies of SOEUs. The result of this study indicates that many of the SOEUs (especially larger ones) are inefficient and perform below the optimal levels. To improve the performance of such SOEUs, policies like restructuring, unbundling, or encouraging competition can be considered.

Keywords Performance evaluation · Relative efficiency · SOEUs · DEA Restructuring

1 Introduction

Efficiency measurement of the electricity industry has always been an important issue. In the present scenario, electric industries of many of the developing countries have been suffered by depriving services and discontent owing to the extensive gap between demand and supply. Due to insufficient revenues and unproductive management, it was very difficult to fill this gap. In case of India, the installed capacity increased from 1713 MW in 1950 to approximately 223,344 MW as on March 2013. The peak shortage in the country for 2013–2014 is 10.42% (Central Electricity

R. V. S. Sengar (✉) · K. Chattterjee
EE Department, Indian Institute of Technology, Dhanbad, Jharkhand, India
e-mail: ramveerchiro@gmail.com

K. Chattterjee
e-mail: kalyanbit@gmail.com

J. Singh
EEE Department, GL Bajaj Institute of Technology and Management, Greater Noida, India
e-mail: jaysinghism@gmail.com

S. N. Singh et al. (eds.), *Advances in Energy and Power Systems*, Lecture Notes in Electrical Engineering 508, https://doi.org/10.1007/978-981-13-0662-4_8

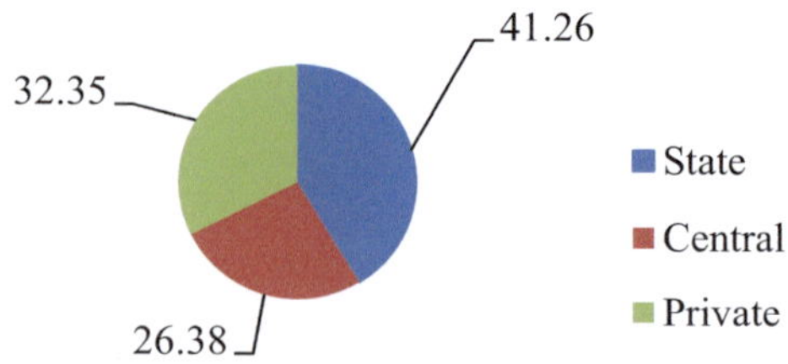

Fig. 1 Installed electrical energy generation capacity sector-wise utilities 31.03.2013

Authority of India, 2014). Installed electrical energy generation capacity sector-wise utilizes as shown in Fig. 1.

To improve the techno-economic efficiency, power sector worldwide is undergoing the process of restructuring and privatization. The purpose of these reforms is to discover market slanting measures for electricity generation and supply it to boost the efficiency of the natural domination behavior of distribution and transmission. In India, several State Electricity Boards (SEBs) are being unbundled into three different corporations, namely generation, transmission, and distribution [1]. Performance evaluation plays a crucial role in structural reforms while defining narrow policies for both distribution and transmission [2]. Benchmarking frameworks for electricity, four distributions have been introduced in Europe and USA [3], and efficiency of South America's electric distribution companies was also estimated by DEA stochastic frontier analysis [4].

For developing countries, several studies are done by different researchers. However, the whole performance analysis of the Indian electrical sector has not been studied, in spite of the very fact that the arena has undergone reforms since 1991, and has further accelerated the method of modification with the Electricity Act, 2005 [5, 6].

There are significant changes within the Indian power sector since early 1990. The number of policies encouraged by SEBs is due to continuous power lack, poor operational performance, and unstable financial state of affairs. These policy changes in relation to institutional and instruction are expected to address the technical and financial challenges faced by Indian power sector [7]. The Indian power sector commenced associate degree epoch of reform since the year 1991 [8] when the gap of the arena for self-employed power producers. With the introduction of Electricity Act, 2003, Indian power sector has been widely reformed in various areas like open access, power trading, and specially in rural area electrification. The act also forces all state to restructure their State Electricity Boards (SEBs). However, these reforms were not adequate to bring about profitable feasibility of SEBs.

2 Literature Review

Data Envelopment Analysis developed by Charnes et al. [9] is used to evaluate the relative efficiency of the generation, transmission, and distribution systems and is implemented by most of the researchers in the electricity sector because of its simplicity with restricted data sets. Data Envelopment Analysis, a nonparametric approach, is based on the linear programming and does not require specification of a production or cost function. It is based on its ability to calculate efficiencies in systems featuring multiple inputs and outputs. Many researchers and economists have applied DEA to evaluate the performance of different utilities under different tenures in the early 1980s.

DEA approach was firstly used by Fare et al. [10] to compare the efficiency of private and public electric utilities producing multiple outputs. Technical efficiency of China's thermal power plants based on the cross-sectional data for 1995 and 1996 was measured by Lam and Shiu [11] using DEA that examines the input slack presence and factors that affect the efficiency. Goto and Tsutsui [12] compared the technical and overall cost efficiencies bilaterally between Japanese and USA electric utilities in 1984–1993 and indicate the reasons of high electricity tariffs. Chitkara [13] evaluated the operational inefficiencies of generating units of all coal-based generating units belonging to NTPC of India over the period 1991–1995 to consider operational performance statistics. Sueyoshi and Goto [14] projected a slack-adjusted DEA approach to examine the operational performance of Japanese electric power generation companies. A fixed benchmarking by Pahwa et al. [2] helps the organizations to develop strategic plans for improvement in their performance and also to compare best among themselves. To investigate the relative operating efficiencies of 40 power units grouped under 8 plants, Cook and Green [15] proposed a two-stage "hierarchy" model. Thakur et al. [8] presented DEA model for calculating the comparative performance of SOEUs involved in the generation, transmission, and distribution of electricity. Wang [16] presented a Performance-Based Regulation (PBR) model to evaluate the total factor productivity with Malmquist productivity index with the help of price cap to analyze the efficiency of Hong Kong's electricity supply industry and its effects on prices. The results support the use of the approach to account for the relation of the PBR model. The contribution of regulatory reforms on productivity growth of Japan's steam power generation sector was measured by Nakano and Managi [17] over the period 1978–2003. Sueyoshi et al. [18] used the DEA to evaluate the performance of US coal-fired power plants. Their outcome suggests that plant managers need to balance between their operational efficiency and environmental performance. A comprehensive study of the Indian power sector reform by Yadav et al. [19] concluded that the sector has considerably improved its operational performance, although no significant improvement is observed in its economical performance. Yadav, et al. [20] proposed a restructuring model using a reorganization algorithm to compare the relative performance of Electricity Distribution Divisions (EDD) of Uttarakhand (an Indian state) using DEA.

3 Methodology

Charnes et al. [9] suggested DEA, a multifactor productivity analysis model for measuring the relative efficiency of a homogeneous set of Decision-Making Units (DMUs). This technique can be used to analyze multiple outputs and multiple inputs without pre-assigned weights and without imposing any functional form on the relationships between variables. Subsequently, a number of models and modification have also evolved. The models proposed in the current study for the investigation are described in the subsequent paragraphs.

$$\text{Max. } \theta_o = \sum_{r=1}^{s} u_r y_{rj} \tag{1}$$

Subject to

$$\left.\begin{array}{l} \sum_{i=1}^{m} v_i y_{ij} = 1 \\ \sum_{r=1}^{s} u_r y_{rj} - \sum_{i=1}^{m} v_i x_{ij} \leq 0; \forall j \\ u_{r,} v_i \geq 0, \forall r \text{ and } i \end{array}\right\} \tag{2}$$

where

θ_o	relative efficiency of DMUs
o	the specific DMU under evaluation
I	the set of inputs ($i = 1, 2, \ldots, m$)
R	the set of outputs($r = 1, 2, \ldots, s$)
J	the set of DMUs, i.e., SOEUs ($j = 1, 2, \ldots, n$)
x_{ij}, y_{rj}	the inputs and outputs of the jth DMU
v_i, u_r	the input and output weights
r	no. of inputs
s and m	are the number of outputs

The above model given in Eqs. (1) and (2) is known as the CCR (Charnes, Cooper, and Rhodes) model in multiplier form based on a Constant Return to Scale (CRS) assumption proposed by Charnes et al. Li and Reeves [21] proposed a model for efficiency evaluation based on deviation variables. If d_o is the deviation variable for DMU_o and d_j is the deviation variable for DMU_j, then, according to Li and Reeves [21], above model can be expressed as:

$$\text{Min. } d_0 \tag{3}$$

Subject to

$$\left.\begin{aligned}&\sum_{i=1}^{m} v_i x_{ij} = 1\\&\sum_{r=1}^{s} u_r y_{rj} - \sum_{i=1}^{m} v_i x_{ij} + d_0 \le 0;\ \forall j\\&u_{r,} v_i \ge 0,\ \forall r \text{ and } i\end{aligned}\right\} \tag{4}$$

In this model, DMU_o is efficient only if $\text{d}_0 = 0$ or $\theta_o = 1$. For inefficient DMU_o, the efficiency score can be calculated as $\theta_o = 1 - \text{d}_0$. For minimizing the sum of deviation, another model is given as in Eqs. (5) and (6):

$$\text{Min.}\, d_j \tag{5}$$

Subject to

$$\left.\begin{aligned}&\sum_{i=1}^{m} v_i x_{ij} = 1\\&\sum_{r=1}^{s} u_r y_{rj} - \sum_{i=1}^{m} v_i x_{ij} + d_j = 0;\ \forall j\\&u_{r,} v_i, d_j \ge 0,\ \forall r,\ i\ and\ j\end{aligned}\right\} \tag{6}$$

The efficiency defined under minimum criteria is more restrictive than defined in conventional DEA; it means that it is rather difficult for DMUs to obtain minimum efficiency frontier as compared to conventional DEA. Minimum criteria yield fewer efficient DMUs.

4 Selection of Inputs and Outputs

Selection of variables, i.e., input and output, is one of the important tasks for performance assessment. The selected variables depend on the several parameters such as methodology, technical requirements of the chosen model, data availability and its quality, and countries own economic structure. However, there is no universal method to choose inputs and outputs. Based on international experience, Jamasb and Pollitt [22] summarize the most widely used variables. Barros and Peypoch [23] also give various methods for selecting inputs and outputs. The input/output selection for the current study was made in concern of those parameters that straight affect the customers in terms of cost of electricity supply. This study included two models to assess the efficiency of SOEUs. In model 1, total cost includes the cost of the Fuel, Operation and Maintenance cost (O&M), Administrative and General cost (A&G), interest payment liability, depreciation and the purchase power expenses. "Adjusted cost" and "number of employees" are the two inputs for Model 2. "Adjusted cost" is the cost component which does exclude the salary and remuneration of the staffs

Table 1 Input and output variables for the two models

S. No.	Model	Input	Output
1	Model 1	Total cost (Rs. millions)	Energy sold (Mkwh), customers (million), distribution line length (circuit km)
2	Model 2	Adjusted cost (Rs. millions), number of employees	Energy sold (Mkwh), customers (million), distribution line length (circuit km)

in the "total cost." Outputs for both the models are length of distribution line, total energy sold and number of consumers, which represents the activity level of the utilities. Table 1 shows the input and output variables chosen for the current work. In both models, an input-oriented approach was chosen to produce a given output at minimal cost. The proposed work consists of the CCR model developed by Charnes which evaluates the overall efficiency score through CRS efficiency and BCC model developed by Banker which evaluates both technical and scale efficiency through Variable Return to Scale (VRS).

5 Results and Discussions

Presently, DEA is an important tool in the area of management science and operation research to evaluate the various managerial and economic problems. The classical DEA has been suffered from weight distributed inputs and outputs. In this work, added programming approach of DEA is employed. Also, 28 Indian SOEUs data have been taken for DEA. Data and results of DEA are given in the subsequent sections.

5.1 Data

For the efficiency comparison of the 28 Indian SOEUs, DEA was used. The physical data for various states were obtained for the year 2013–2012, from general review (2013–2014) published by CEA [24]. The cost data for the same time period were obtained from Annual Report (2014) [25], published by the planning commission, Government of India. A summary of the data is presented in Table 2, in the form of mean, median, standard deviation, maximum and minimum values. The main difference comes into view to be present among various utilities as is evident from the last two columns of Table 2. The values of correlation coefficients (Table 3) indicate that the variables are sensibly correlated [26]. Figure 2 shows the average breakup of the unit cost of supply. From Fig. 2, it is clear that the power purchase cost (approximately 75.5%) is the largest component and then A&G expenses (8.75%)

Table 2 Summary statistics for the data

Variable	Mean	Median	Standard deviation	Minimum	Maximum
Energy sold (10^6 *kwh)	26516.3	9915.8	29265.6	415.3	96461.7
No. of consumers (million)	7.23214	3.8	7.88322	0.1	24.7
Distribution line length (km)	283,932	166,775	304,157	4883	903,408
Total cost (Rs. millions)	157,276	80555.6	174,725	2793.05	563,520

Table 3 Input/output correlations

Variable	Energy sold (10^6 *kwh)	Number of consumers (million)	Distribution line length (km)	Total cost (Rs. million)
Energy sold (10^6 *kwh)	1	0.929936009	0.94025157	0.987602687
Number of consumers (million)	0.929936009	1	0.950935257	0.921301784
Distribution line length (km)	0.94025157	0.950935257	1	0.916262065
Total cost (Rs. million)	0.987602687	0.921301784	0.916262065	1

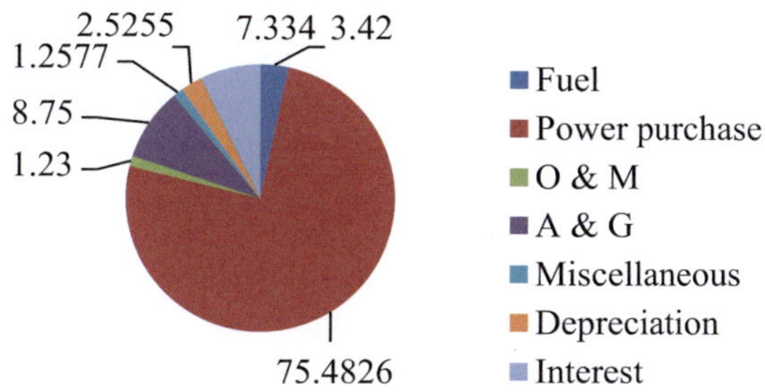

Fig. 2 Components of unit supply cost. *Source* Annual Report, 2013

and then fuel costs (approximately 7.3%). Among the above costs, A&G cost can be controlled by decision-makers because it comprises mainly salaries and wages of the employees. That is why in Model 2 "number of employees" is taken as additional input along with "adjusted cost" for the efficiency calculation using DEA.

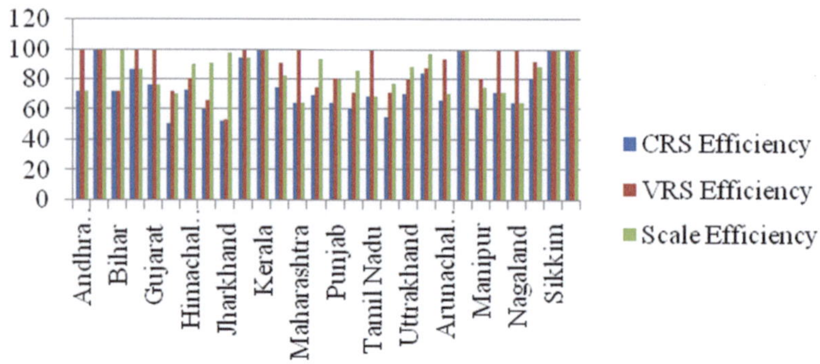

Fig. 3 Various types of efficiency

5.2 *Model 1 (Input: Total Cost)*

5.2.1 CCR Formulation

For the case of electric utilities, several studies have established that the CCR model is most suitable. The CCR model generates CRS efficiency and does not distinguish between units based on their sizes in the computation of their efficiencies. The results of this model are presented in Table 4. It is clear from Table 4 that SOEUs show significant variations in efficiency levels. The total efficiency had a mean score of 75% for all the utilities, and most of the utilities (17 out of 28) lie below this mean value. Five utilities turned out to be the best performers, and the remaining utilities exhibited varying degrees of inefficiencies.

5.2.2 BCC Formulation

BCC formulation compared and benchmarked the similar-sized utilities because it assumes a VRS by taking into consideration the size of the utilities was employed. The results are presented in Table 5. Now, the efficient utilities increased to 13, while the other utilities showed inefficiencies of varying degrees. Figure 3 shows the various types of efficiencies for each utility. It is clear from Table 5 that Andhra Pradesh, Maharashtra, Mizoram, Nagaland, and some more SOEUs are indicating the potential for efficiency enhancement by suitable restructuring. The cost saving using the Model 1 (BCC formulation) is also presented in Table 5. Table 5 established that theoretically it is possible to save Rs. 17,370 million per year with respect to present best practices in the SOEUs. Actual saving may be quite lesser.

Table 4 Result of model 1 (CCR formulation)

S. No.	State	CCR efficiency %	Efficient	Benchmark
1	Andhra Pradesh	73		11, 27, 28
2	Assam	100	Yes	2
3	Bihar	73		11, 27
4	Chhattisgarh	86		2, 27
5	Gujarat	77		11, 27, 28
6	Haryana	51		11, 27
7	Himachal Pradesh	73		11, 27
8	Jammu and Kashmir	60		2, 27
9	Jharkhand	53		11, 27
10	Karnataka	95		11, 27, 28
11	Kerala	100	Yes	2, 11
12	Madhya Pradesh	75		2, 27
13	Maharashtra	64		11, 27, 28
14	Meghalaya	70		2, 27
15	Punjab	64		11, 27, 28
16	Rajasthan	61		2, 27
17	Tamil Nadu	69		11, 27
18	Uttar Pradesh	55		11, 27
19	Uttarakhand	71		2, 27, 28
20	West Bengal	84		11, 27, 28
21	Arunachal Pradesh	66		2, 27
22	Goa	100	Yes	11, 27
23	Manipur	60		2, 27
24	Mizoram	71		11, 27, 28
25	Nagaland	65		2, 27
26	Pondicherry	81		11, 27
27	Sikkim	100	Yes	27
28	Tripura	100	Yes	28
Average CCR efficiency % = 75				

5.3 Model 2: (Input: Adjusted Cost and Number of Employees)

To evaluate the possibility of reduction of employees, CCR and BCC models were run with the number of employees acting as an additional input. The other input was “adjusted cost” (total cost adjusted for personnel cost). Table 6 presents the results of Model 2, which present the various efficiencies along with benchmarks and return

Table 5 Results of model 1 (BCC formulation)

S. No.	State	Technical efficiency (%)	Scale efficiency (%)	Savings (Rs millions)
1	Andhra Pradesh	100	73	0
2	Assam	100	100	0
3	Bihar	73	100	21316.5414
4	Chhattisgarh	100	86	0
5	Gujarat	100	77	0
6	Haryana	72	71	72733.17638
7	Himachal Pradesh	81	91	9146.482804
8	Jammu and Kashmir	66	91	14835.0855
9	Jharkhand	54	98	37794.04931
10	Karnataka	100	95	0
11	Kerala	100	100	0
12	Madhya Pradesh	91	82	23313.95468
13	Maharashtra	100	64	0
14	Meghalaya	75	93	2476.535655
15	Punjab	81	80	43865.90072
16	Rajasthan	71	86	96004.81578
17	Tamil Nadu	100	69	0
18	Uttar Pradesh	71	77	130426.6411
19	Uttarakhand	80	88	9968.2876
20	West Bengal	87	97	22236.54029
21	Arunachal Pradesh	94	71	260.4605238
22	Goa	100	100	0
23	Manipur	81	75	989.1916743
24	Mizoram	100	71	0
25	Nagaland	100	65	0
26	Pondicherry	92	88	977.0260032
27	Sikkim	100	100	0
28	Tripura	100	100	0
	Average	88	85	17369.453

to scale nature. Table 6 indicated the possibility of percentage saving in adjusted cost and number of employees. Theoretically, it is possible to reduce the employees of the states, Jammu and Kashmir, Haryana Rajasthan, Uttar Pradesh, and Manipur. The overall possibility of staff reduction is approximately 36,171 employees (5.33% of total employees) of various utilities. By the curtailment of staff, it is possible to improve the efficiency of various SOEUs shown in Table 6.

Table 6 Results of model 2

S. No.	Utility	Total efficiency (%)	Efficient	Benchmark CRS	Technical efficiency (%)	Benchmark VRS	Returns to scale	Scale efficiency
1	Andhra Pradesh	98		19, 20	100	1	Decreasing	98
2	Assam	100	Yes	2	100	2	Constant	100
3	Bihar	82		11, 20	84	11, 20, 24	Increasing	97
4	Chhattisgarh	100	Yes	4	100	4	Constant	100
5	Gujarat	81	Yes	4, 11, 22	100	5	Decreasing	81
6	Haryana	65		4, 19	75	1, 4, 5	Decreasing	85
7	Himachal Pradesh	85		2, 4, 22	94	4,11, 22	Decreasing	90
8	Jammu and Kashmir	65		2, 4, 22	68	2, 4, 22	Decreasing	95
9	Jharkhand	86		19, 20	97	20, 26	Increasing	89
10	Karnataka	100	Yes	10	100	10	Constant	100
11	Kerala	100	Yes	11	100	11	Constant	100
12	Madhya Pradesh	87		2, 4, 22	98	4, 5, 10	Decreasing	88
13	Maharashtra	81		4, 19, 20	100	13	Decreasing	81
14	Meghalaya	86		2, 4, 22	93	4, 22, 24, 28	Increasing	92
15	Punjab	79		2, 4, 11, 22	93	4, 5, 10	Decreasing	85
16	Rajasthan	77		4, 10, 20	81	1, 10,20	Decreasing	95
17	Tamil Nadu	71	Yes	4,10,11	100	17	Decreasing	71
18	Uttar Pradesh	80		19, 20	86	1,4	Decreasing	93
19	Uttarakhand	100	Yes	19	100	19	Constant	100
20	West Bengal	100	Yes	20	100	20	Constant	100
21	Arunachal Pradesh	80	Yes	2, 27	100	21	Increasing	80
22	Goa	100	Yes	22	100	22	Constant	100
23	Manipur	65		2,27	81	2, 21, 24	Increasing	79
24	Mizoram	75	Yes	2, 11, 27	100	24	Increasing	75
25	Nagaland	67		2, 27	97	2, 21, 24	Increasing	69
26	Pondicherry	96	Yes	4, 22	100	26	Increasing	96
27	Sikkim	100	Yes	27	100	27	Constant	100
28	Tripura	98	Yes	2, 11, 22, 27	100	28	Increasing	98

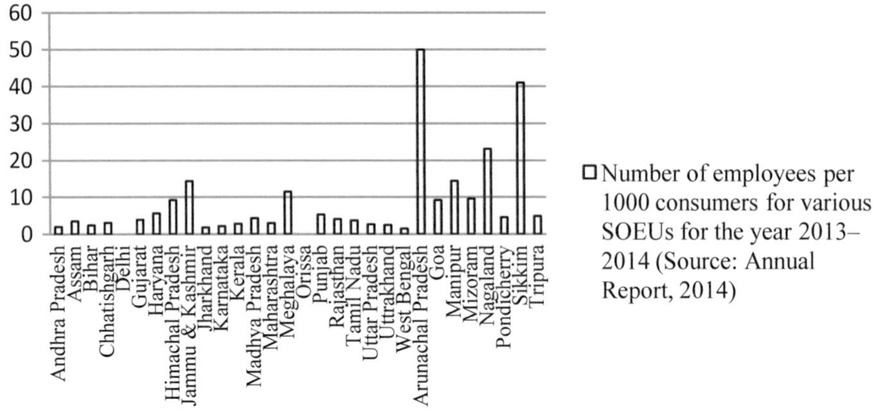

Fig. 4 Number of employees per 1000 consumers for various SOEUs for the year 2013–2014. *Source* Annual Report, 2014

In Fig. 4, a wide variation of employees per 1000 consumers (Staff Productivity Index) has been illustrated. Many of the researchers already indicated that many of the Indian SOEUs suffer from over-employment [27, 28]. For some utilities like Arunachal Pradesh, Sikkim, and Nagaland, this ratio is too large (Table 7).

Table 7 Possible savings (VRS) in Model 2

S. No.	Utility	Adjusted cost (Rs. million)	Number of employees	Savings		Percentage savings	
				Adjusted cost (Rs. million)	Number of employees	Adjusted cost%	Number of employees (%)
1	Andhra Pradesh	470044.44	47,303	0	0	0	0
2	Assam	29008.61	10,382	0	0	0	0
3	Bihar	72630.84	13,500	11410.75	2121	15.7	15.7
4	Chhattisgarh	90313.52	12,000	0	0	0	0
5	Gujarat	294825.44	54,102	0	0	0	0
6	Haryana	243461.42	30,623	59764.25	7517	24.5	24.5
7	Himachal Pradesh	36072.734	19,300	2267.511	1213	6.3	6.3
8	Jammu and Kashmir	37105.50	18,673	11831.144	5954	31.9	31.9
9	Jharkhand	76982.54	6650	2625.417	227	3.4	3.4
10	Karnataka	235352.82	42,698	0	0	0	0
11	Kerala	82008.66	31,019	0	0	0	0

(continued)

Table 7 (continued)

S. No.	Utility	Adjusted cost (Rs. million)	Number of employees	Savings		Percentage savings	
				Adjusted cost (Rs. million)	Number of employees	Adjusted cost%	Number of employees (%)
12	Madhya Pradesh	222083.01	47,496	4071.3989	871	1.8	1.8
13	Maharashtra	520603.79	61,276	0	0	0	0
14	Meghalaya	7739.67	3450	513.1156	229	6.6	6.6
15	Punjab	185973.28	44,132	12292.892	2917	6.6	6.6
16	Rajasthan	313002.12	45,887	60538.813	8875	19.3	19.3
17	Tamil Nadu	422480.41	92,014	0	0	0	0
18	Uttar Pradesh	428865.44	39,546	60206.595	5552	14	14
19	Uttarakhand	46928.32	4427	0	0	0	0
20	West Bengal	160319.72	19,209	0	0	0	0
21	Arunachal Pradesh	3069.34	10,000	0	0	0	0
22	Goa	9059.029	5497	0	0	0	0
23	Manipur	4163.447	2892	781.5295	543	18.8	18.8
24	Mizoram	2295.529	1919	0	0	0	0
25	Nagaland	3192.2008	4621	105.6387	153	3.3	3.3
26	Pondicherry	11556.719	1837	0	0	0	0
27	Sikkim	2494.6065	4100	0	0	0	0
28	Tripura	6749.9685	3449	0	0	0	0
Total		4018383.2	678,002	226409.07	36,171		

6 Conclusions

This study proposes an enhanced nonparametric approach of DEA to calculate the efficiency of 28 Indian electric utilities and provide the possible targets for their improvement. This study makes an attempt to present the scrutiny enhancement in the effective performance of the SOEUs in the Indian power. The result of this study indicates that many of the SOEUs are inefficient and perform below the optimal levels. Most of the inefficient SOEUs are suffered from scale inefficiency as compared to technical efficiency. The result also indicates the cost saving by the curtailment of number of staff; approximately 5% employees can be curtailed. The extent of over staffing may assume critical implication when efficacy reformation takes place, as it is not easy to cut back staff. Still, there are quite a lot of potential improvements which need the attention of the policy makers if the power sector of the country aims at excellence. To improve the performance of SOEUs, policies like restructuring, unbundling, or encouraging competitions can be considered.

Glossary

A&G	Administrative and General
BCC	Banker, Charnes, and Cooper
CCR	Charnes, Cooper, and Rhodes
CEA	Central Electricity Authority
CRS	Constant Return to Scale
DEA	Data Envelopment Analysis
DMUs	Decision-Making Units
EDDs	Electricity Distribution Divisions
NTPC	National Thermal Power Corporation
O&M	Operating and Maintenance
PBR	Performance-Based Regulation
SEBs	State Electricity Boards
SOEUs	State Owned Electric Utilities
VRS	Variable Return to Scale

References

1. Singh, S.N., Srivastava, S.C.: Electric power industry restructuring in India: present scenario and future prospects. In IEEE Conference (DRPT), pp. 20–23 (2004)
2. Pahwa, A., Feng, X., Lubkeman, D.: Performance evaluation of electric distribution utilities based on data envelopment analysis. IEEE Trans. Power Syst. **17**(3) (2002)
3. Burns, P., Weyman-Jones, T.: Cost functions and cost efficiency in electricity distribution: a stochastic frontier approach. Bull. of Econ. Res. **48**(1), 41–64 (1996)
4. Estache, A., Rossi, M.A., Ruzzier, C.A.: The case for international coordination of electricity regulation: evidence from the measurement of efficiency in South America Washington, D. C. J. Regul. Econ. **25**(3), 271–295 (2004)
5. Thakur, T., Deshmukh, S.G., Kaushik, S.C., Kulshrestha, M.: Impact assessment of the electricity act 2003 on the Indian power sector. Energy Policy **33**(9), 1187–1198 (2006)
6. Sharma, D.P., Chandramohanan Nair, P.S., Balasubramanian, R.: Performance of Indian power sector during a decade under restructuring: a critique. Energy Policy **33**, 563–576 (2005)
7. Singh, A.: Power sector reform in India current issues and prospects. Energy Policy **34**, 2480–2490 ((2006))
8. Thakur, T., Deshmukh, S.G., Kushik, S.C.: Efficiency evaluation of the state owned electric utilities in India. Energy Policy **34**, 2788–2804 (2006)
9. Charnes, A., Cooper, W.W., Rhodes, E.: Measuring the efficiency of decision making units. Eur. J. Oper. Res. **2**, 429–444 (1978)
10. Fare, R., Grosskopf, S., Lovell, C.A.K., Pasurka, C.: Multi lateral productivity comparisons when some outputs are undesirable: a nonparametric approach. Rev. Econ. Stat. **71**, 90–98 (1989)
11. Lam, P.L., Shiu, A.: A data envelopment analysis of the efficiency of China's thermal power generation. Util. Policy **10**, 75–83 (2001)
12. Goto, M., Tsutsui, M.: Comparison of productive and cost efficiencies among Japanese and US electric utilities. Omega **26**, 177–194 (1998)
13. Chitkara, P.: A data envelopment analysis approach to evaluation of operational inefficiencies in power generating units: a case study of Indian power plants. IEEE Trans. Power Syst. **12**(2), 419–425 (1999)

14. Sueyoshi, T., Goto, M.: Slack-adjusted DEA for time serious analysis: performance measurement of Japanese electric power generation industry in 1984–199. Eur. J. Oper. Res. **133**, 232–259 (2001)
15. Cook, W.D., Green, R.H.: Evaluating power plant efficiency: a hierarchical model. Comput. Oper. Res. **32**, 813–823 (2005)
16. Wang, J.H., et al.: Performance based regulation of the electricity supply industry in Hong Kong: an empirical efficiency analysis approach. Energy Policy **35**, 609–615 (2007)
17. Nakano, M., Managi, S.: Regulatory reforms and productivity: an empirical analysis of the Japanese electricity industry. Energy Policy **36**, 201–209 (2008)
18. Sueyoshi, T., Goto, M., Ueno, T.: Performance analysis of US coal-fired power plants by measuring three DEA efficiencies. Energy Policy **38**, 1675–1688 (2010)
19. Yadav, V.K., Padhy, N.P., Gupta, H.O.: A micro level study of an Indian electric utility for efficiency enhancement. Energy **35**(10), 4053–4063 (2010)
20. Yadav, V.K., Chauhan, Y.K., Padhy, N.P., Gupta, H.O.: A novel power system restructuring model based on data envelopment analysis (DEA). Electr. Power Energy Syst. **44**, 629–637 (2013)
21. Li, X.B., Reeves, G.R.: A multi criteria approach to data envelopment analysis. Eur. J. Oper. Res. **115**, 507–517 (1999)
22. Jamasb, T., Pollitt, M.: International benchmarking and regulation: an application to European electricity distribution utilities. Energy Policy **31**, 1609–1622 (2003)
23. Barros, C.P., Peypoch, N.: Technical efficiency of thermoelectric power plants. Energy Econ. **30**(1), 3118–3127 (2008)
24. General Review 2013–2014. All India Electricity Statistics, Central Electricity Authority, Ministry of power, India (2014)
25. Ministry of Power, Government of India (2014). Available: http://powermin.nic.in/
26. Golany, B., Roll, Y.: An application procedure for DEA. Omega **17**(3), 237–250 (1989)
27. Kannan, K.P., Pillai, N. V.: The aetiology of the inefficiency syndrome in the Indian power sector. Working Paper No. 324 Centre for Development Studies, Thiruvananthapuram (2002)
28. Gurtoo, A.: North electricity board: a case of restructuring in the Indian power industry. Asian Case Res. J. **6**(1), 27–43 (2002)

A Novel Scheme of Fault Detection in Transmission Line Using Image Processing

Deepak Kumar, Amit Kumar, Abhay Yadav and M. A. Ansari

Abstract This paper gives a scheme to find the fault or monitor the condition of transmission line using image processing. The technique of Digital Image Processing (DIP) and Wavelet Shrinkage Function (WSF) is used for fault detection and diagnosis. In other context, image of a transmission line is taken by the thermovision camera with the coordinates of transmission line, which are current, voltage, and temperature. The algorithm for image segmentation is used which divides the image into the set of part and objects. Application of WSF is done to read the image characteristics and standard deviation which gives the image quality where the fault occurs. This proposed method gives results in terms of visual quality and peak signal-to-noise ratio (PSNR).

Keywords Image processing · Fault detection · Transmission line · Wavelet function

1 Introduction

With the use of image processing system, the image can be converted into the digital form and useful information can be extracted by using some mathematical operation. Noise is the main problem which reduces the image quality that causes error shown in the image. So de-noising is an important task in the image processing, and the wavelet function is the best function to reduce the noise from the image. The wavelet

---

D. Kumar (✉) · A. Kumar · A. Yadav · M. A. Ansari
School of Engineering, Gutam Buddha University, Greater Noida, India
e-mail: deepu1796@gmail.com

A. Kumar
e-mail: abes.amit@gmail.com

A. Yadav
e-mail: abhayadav@outlook.com

M. A. Ansari
e-mail: mahmadiitr@gmail.com

S. N. Singh et al. (eds.), *Advances in Energy and Power Systems*, Lecture Notes in Electrical Engineering 508, https://doi.org/10.1007/978-981-13-0662-4_9

transform function is applied to the image at different points for noise reduction. During the acquisition process, the noise is occurred in images, due to intrinsic and thermal fluctuation of source device. Image processing is also used in the medical field of the essence. During the disease diagnosis, many types of equipment that are used in the medical field are in digital format. For fault detection, the image taken by the thermovision camera is very helpful in finding the fault. Hence, first data is collected and then segmentation processes are applied which followed wavelet function to calculate the standard deviation. This is used to develop diagnosis tool known as bagging strategy which raises the generalization power of fault system. This technique is very helpful, to solve the problems in all areas of power system like heat detection in MCB switch or for the arrester heat.

DIP has the set of functions that make transfiguration of the images to improve the information from machines [1]. The functions are utilized to include the most ideal number of neuro-fuzzy systems in the thermovision analysis [2]. Methods related to fault detection on transmission line using neural network techniques are proposed. Dengwen and Wengang [3] proposed discrete wavelet transform (DWT) for image de-noising and gave performance in terms of peak signal-to-noise ratio. Fan et al. [4] proposed set partitioning in hierarchical trees (SPIHT) method to reduce the distortion in images. Pan [5] presented a remote monitoring of transmission lines using image processing. Jun and Xinyu [6] presented thermal image processing, which is very useful in power station to detect the heating range of electrical equipment and read the image pixels that contain the higher value than other pixels of image.

Chinnarao and Madhavilatha [7] use the methodology to reduce the noise from the image without affecting important features of image and content, and use the dual tree complex wavelet transform function for thresholding between noisy feature and important feature of image. Kumar et al. [8] provide better technique to reduce the noise from the image and enhance the image quality and give better views.

2 Process of Fault Detection and Diagnosis

This can be divided into the following steps which are shown in Fig. 1.

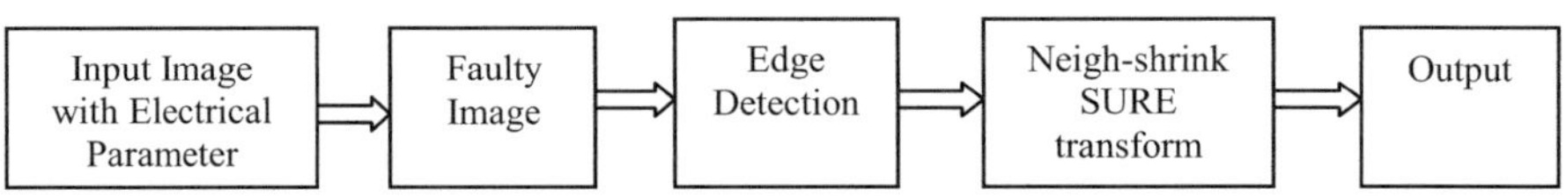

Fig. 1 Process of identification of fault

Table 1 Imaging performance

Spatial resolution	1.3 mrad.
Digital image enhancement	Normal and enhanced
Detector type	Focal plane array uncooled microbolometer
Spectral range	8 to 14 μm
Focus	Automatic or manual
Thermal sensitivity @ 50/60 Hz	0.08 °C at 36 °C

2.1 Image Acquisition

Firstly, we will take an image by thermovision or charge-coupled device camera for recording purpose and then convert it into digital form. After the image obtained, various function are to be applied to the image for the fault analysis and diagnosis. The parameters of the image are shown in Table 1.

2.2 Wireless Sensors

A wireless sensor network (WSN) is a wireless network which consists of sensors. The charge-coupled device cameras are widely used in field of wireless sensor network. Sensors are used to monitor physical and environmental conditions. The image will be taken from the transmission end and transmitted to receiving end terminal for the analysis of image.

2.3 Image Segmentation and Edge Detection

The image segmentation is the process where the object is located and border is marked. By this process, the image gets separated from the other parts of the image. Fault identification technique is applied to this separated part from Image. The image has entire data that have temperature, voltage, current, of the symmetrical transmission as input parameter and the image is set for transmission. The most useable functions utilized for the division are point or line discovery, edge identification, gradient administrators, Laplacian, Hough light, simple or adaptive threshold, district developing and watershed change [5]. This paper uses edge detection which is the part of process called segmentation method for better edge preservation. Edge detection is the most popular operations for image analysis as images have some features like line, edge, points so the detection of edge can be possible. The common algorithms used for edge detection are Sobel, Canny,

Prewitt, and gradient operators. Sobel and Prewitt are also known as directional edge detectors so these edge two operations are mostly used.

2.4 Image De-noising

In the power system, the fault occurs in the transmission line and faulty image which is taken by the source. The image has the noisy value which shows that the fault occurs in the line and noisy value is directly related to the voltage, current, and temperature level. The faulty images will get reclaimed to the initial values by thresholding the values. The principle is that when noise occurs as fault during transmission, it will be sent to the other node. The recapture of the fault can be done by wavelet shrinkage function.

2.5 Wavelet Shrinkage Function

Neigh-Shrink SURE transformation is one of the ways in wavelet shrinkage function, used to spot the faults in transmission line. This is best method for image noise reduction. A non-excess, orthogonal, wavelet transform is being connected to the corrupt data, taken by vectoresteemed thresholding of individual multichannel wavelet coefficient that square measure at long last conveyed back to the picture area by converse wavelet change. The Neigh-Shrink SURE is useful algorithm for image de-noising. The Neigh-Shrink, threshold, and neighboring window size can be determined by the following algorithm as:

$$(P^{s}, Q^{s}) = \arg_{p,q} \min \mathrm{SURE}(W_s, P, Q) \tag{1}$$

where P is optimal threshold, Q is neighboring window size, s is sub-band and SURE.

$$(w_s, P, Q) = N_s + \Sigma \, ||g_n(w_n)||_2^2 + 2 \, \Sigma \, (\partial g_n / w_n) \tag{2}$$

$\mathrm{N_s}$: noisy wavelet coefficients from sub-band s

$W_s = \{W_{jk} : \; j, k \in \text{ Indices corresponding sub - bands } s\}$

Into the 1-D vector

$$g_n(w_n) = \begin{cases} -\frac{p^2}{S_n^2} W_n(P\, S_n) \\ -W_n(\text{otherwise}) \end{cases} \tag{3}$$

In this Eq. (3) $g_n(w_n)$ is wavelet coefficient.

Image thresholding is a simple, useful, way of separating an image into a foreground and background. This image analysis mode is a type of image segmentation that isolates items by transforming grayscale images into binary images. Image thresholding is most useful in images with high levels of contrast. This technique specifies an edge esteem for every level in wavelet transform to indicate the level of universal thresholding. It means the portion of image where the contrast value is high, the high contrast value means fault is in this portion and the standard value of this portion is very low. If the part of the image has a good standard value that means no fault occur and if the threshold and standard value is high it means fault occurs in the line. It is characterized as [3, 4].

$$g(x, y) = \begin{cases} 1, & f(x, y) > T_d \\ 0, & f(x, y) \le T_d \end{cases} \quad (4)$$

where T_d is threshold, and the image background and object pixels replace by 0 and 1.

$$T = \sigma\sqrt{2 \log M} \quad (5)$$

where σ is standard deviation which is used for the wavelet coefficient sub-band

$$\sigma = \frac{\text{median}(|w_s|)}{0.6745}(w_s \in \text{subband}) \quad (6)$$

In all wavelet sub-band s, Neigh-Shrink used the universal threshold $\lambda = \sigma\sqrt{2\log(512)} \approx 3.23\sigma$

The limit and neighboring window size of the proposed technique in each sub-band were computed with Eq. (1).

3 Results and Discussion

In order to estimate the execution of the proposed method, set of 8-bit standard grayscale Computer Vision Group CVG-UGR database are used, such as house, Barbara, pepper, mandrill each of size 512 × 512, 256 × 256, which are corrupted by simulated additive white Gaussian noise with a standard deviation equal to 5, 10, 15, 20. The noisy image is filtered by these several techniques. Calculate the PSNR values as follows:

$$\text{PSNR} = 10 * \log_{10} \frac{255^2}{\text{MSE}}(dB) \quad (6)$$

The performance of this method is compared with other schemes, where mean square error (MSE) is defined as

Fig. 2 Input image as gray image

Fig. 3 Faulty image

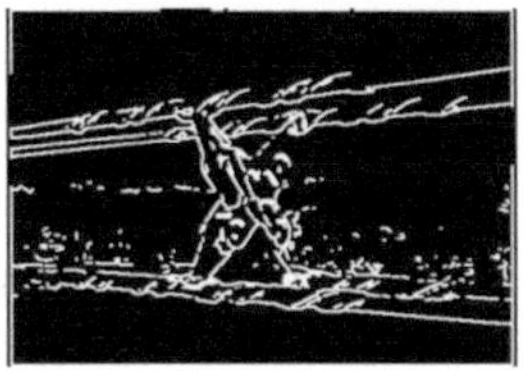

Fig. 4 Dilated gradient mask image

$$MSE = \frac{1}{mn} \sum_{x=0}^{m-1} \sum_{y=0}^{n-1} \|I(x, y) - f(x, y)\|^2 \tag{7}$$

By these equations, calculate the peak signal-to-noise ratio value and the mean square error with the elapsed time. The image takes from the source, applied edge detection and Neigh-Shrink SURE transformation. The results given by the edge detection segmentation process images are shown below as Fig. (2) show input image, Fig. (3) show Faulty image, Fig. (4) show dilated gradient image and Fig. (5) show diagnosis image.

- Figure 2, shows the image which is taken by the source and then apply process.
- Figure 3, is faulty image where the noise level is too high, separate the image by applying segmentation process with input parameter. The image shows in Fig. 4. After that point out the joint where the transmission line is jointed with the supporters because most of fault occur by the loose connection.
- Taking loose connection point and find out noise level which is directly related to standard deviation if noise increase, the value of standard deviation is also increase so chances of fault will increases.
- Figure 5, shows the diagnosis image which is after the fault (Table 2).

Fig. 5 Diagnosis image

Table 2 Calculated PSNR value and elapsed time

Input image	Noise level	PSNR	MSE	Elapsed time
 Ddss.png	5	38.0733	10.1334	12.035249
	10	34.7204	21.9303	14.353844
	15	32.8676	33.5985	14.500361
	20	31.5302	45.7144	13.737332
	30	29.7023	63.6393	15.089631
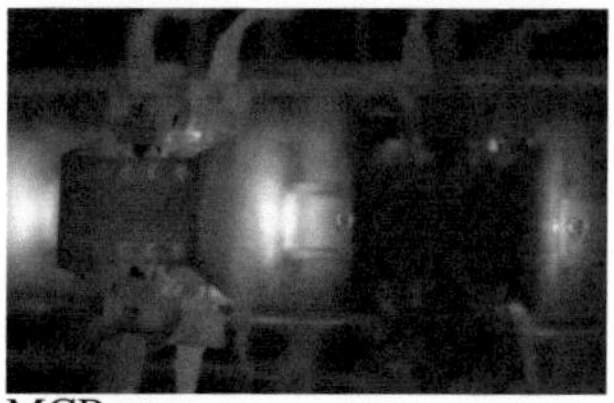 MCB.png	5	37.2030	12.3817	12.599801
	10	33.0243	32.4070	10.604030
	15	30.6878	55.5010	15.688973
	20	29.0935	80.1185	12.402707
	30	27.0058	129.5683	11.490505

4 Conclusion

The input parameter of the transmission line are injected in the image and transmitted to the network and processed by the wavelet shrinkage function. If any fault or any noise occurs in the image of transmission line or in micro circuit board switch, the image changes its characteristics which are proceeding by the Neigh-Shrink SURE function (NSSF) and the original value of the image input can be obtained. Thus, it assured that the location fault can be identified and diagnosed. The diagnostics tool completely showed itself as a powerful tool to identify fault. The operation process speed is very fast, reliability about 95–98%, human factor 5%.

References

1. Om, H., Biswas, M.: A generlized image denoising method using neighbouring wavelet coefficients. Sig. Image Video Process (SIViP) (2013) (Springer)
2. Almeida, C.A.L.: Intelligent thermographic diagnostic applied to surge arresters: a new approach. IEEE Trans. Power Deliv. **24** (2009)
3. Dengwen, Z., Wengang, C.: Image denoising with an optimal threshold and neighbouring window. Pattern Recogn. Lett. **29**(11), 1694–1697 (2008) (Elsevier)
4. Fan, W., Chen, J., Zhen, J.: SPIHT Algorithm Based on Fast Lifting Wavelet Transform in Image Compression. Springer publication (2005)
5. Pan, L.: Intelligent image recognition research on status of power transmission lines. Sens. Transducers **179**, 174–179 (2014) (IFSA publication)
6. Jun, L., Xinyu, L.: Heating defect detection system scheme design based on infrared image processing for high voltage plant of substation. Advance in Control Engineering and Information Science, Elsevier (2011)
7. Chinnarao, B., Madhavilatha, M.: Improved image de noising algorithm using dual tree complex wavelet transform. Int. J. Comput. Appl. **44** (2012)
8. Kumar, D., Kumar, A., Yadav, A.: A flexible scheme of fault sensing in power transmission line using artificial intelligence technics, Am. J. Remote Sens. **4**(6), pp. 33–39 (2016). https://doi.org/10.11648/j.ajrs.20160406.11

Power Quality Improvement of Power Distribution System Under Symmetrical and Unsymmetrical Faults Using D-STATCOM

Amit Kumar, Deepak Kumar and Abhay Yadav

Abstract Voltage and current waveform distortion is a most important power quality concern, because a small fluctuation with phase supply, transient condition occurs and unbalanced of phase voltage can create a more derange within phase currents. Three phase balanced system holds only positive sequence constituents of voltage, current, and impedance, whereas both positive and negative sequence constituents of voltages and currents within unbalanced system. The voltages at the generation system side are in sinusoidal wave and have constant equivalent magnitude value within 120° phase difference. However, unbalanced condition arises due to inadequacy in voltage magnitude and in fundamental frequency in load side. This paper proposes the shunt-connected current inserting Distribution Static Synchronous Compensator (D-STATCOM) through appropriate controlling scheme to moderate and compensate the unstable load current. It works as quick reimbursing source for reactive power that is functional on the distribution power system to moderate voltage deviations like voltage sags, swells, and voltage flicker along with unsteadiness produced by quickly changing reactive power demand. This paper presents the scheme and Simulink model of D-STATCOM within PI controller toward improving the superiority of power under dissimilar irregular circumstances like symmetrical and unsymmetrical faults. The controller working efficiency is tested under different fault conditions. The performance of D-STATCOM under different fault conditions in power distribution system is carried out at three phase supply 20 kV, 50 Hz. The controller and D-STATCOM model are designed through Simulink and Power System Blockset toolbox existing in MATLAB program.

Keywords Power quality · D-STATCOM · Voltage source converter flexible alternating current transmission system · Pulse width modulation

A. Kumar (✉) · D. Kumar · A. Yadav
School of Engineering, Gautam Buddha University, Greater Noida, India
e-mail: abes.amit@gmail.com

D. Kumar
e-mail: deepu1796@gmail.com

A. Yadav
e-mail: abhayadav@outlook.com

S. N. Singh et al. (eds.), *Advances in Energy and Power Systems*, Lecture Notes in Electrical Engineering 508, https://doi.org/10.1007/978-981-13-0662-4_10

1 Introduction

Power quality (PQ) is an important and serious issue in power distribution network system. In present day, the industrial equipment is primarily based on electronic components such as rectifier, thyristors together with programmable sensing controllers and power electronic drives [1]. These devices are very sensitive to instabilities and end up less tolerant to PQ issues [3]. In the industrial and commercial equipment, voltage sags, swell, voltage fluctuation are serious and most common problems [1].

D-STATCOM is a reactive power shunt jointed reimbursement device that is capable for generating and absorbing the reactive power and also provides voltage sustenance to a system bus [4]. In a renewable energy system, it is also utilized in improving the constancy of the power system [2]. The reactive power imbalances can be minimized through the compensation device [5]. D-STATCOM is also capable to solve fault problems and compensate harmonics through capability of the system [7].

The first D-STATCOM was introduced by Japan country and it transported to the UK, where it has been installed in southwest England to adjacent a wind farm.

The Flexible Alternating Current Transmission System (FACTS) offers a fast and consistent control mechanism over transmission line constraints, like voltage, phase angle between the sending end and receiving end voltage, line impedance [8]. Custom power devices (CPD) are designed for distribution of voltage supplies at low voltage and encroachment of PQ due to which the system becomes reliable [11]. Solid-state CPD is utilized in electrical power network to offer reimbursement and advance voltage profile [1]. Some CPD are D-STATCOM, unified power quality conditioner (UPQC) as a combination of series and parallel [6], Dynamic Voltage Restorer (DVR) is in series connection [9] among D-STATCOM which is a cost operational solution for reimbursement of reactive power [6]. A FACTS is a power semiconductor device which is utilized to sustain the PQ by preserving superior movement of power and controlling the dynamically stability of system through altering the system constraints like voltage, angle of phase, and impedance [13].

2 Configuration and Operation of D-STATCOM

A D-STATCOM is voltage source inverter (VSI)-fed semiconductor-based power electronic expedient [2] that is associated with shunt to the system to moderating harmonics and PQ problems [9]. Its performance depends upon different control algorithms, for taking out reference current to deliver pulses to gate-terminals of VSI [10]. The D-STATCOM is extremely effective to providing regulation at load voltage. It keeps rated voltage value at load point that has numerous undesirable things from customer point of view. Within voltage value of 1 per unit at load point, D-STATCOM force to load to drive continuously at rated value [3]. It can easily convert both active and reactive power within electrical power

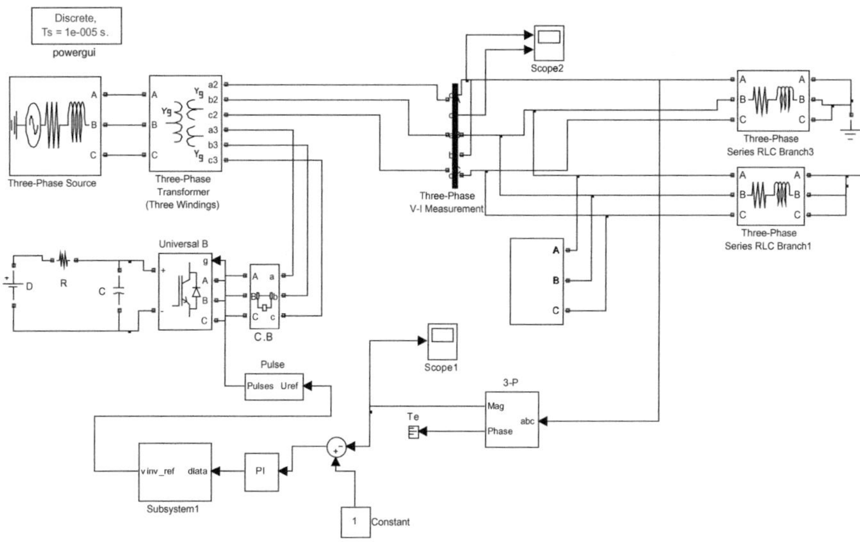

Fig. 1 MATLAB simulink model of D-STATCOM

distribution network through flexible amplitude and phase angle of the voltage converter device [4].

2.1 MATLAB Simulink Model of D-STATCOM

In Fig. 1, three phase supply 20 kV is utilized in power distribution system and nonlinear load is connected as a load in power system network. Three phase distribution transformer is connected with the three phase supply voltage. D-STATCOM is a shunt-connected device which is in between the power supply and nonlinear load. A fault occurs in distribution line near a load in a transmission line network. Circuit breaker is used to isolate the faulty network in power system network which can observe the difference at the time of fault and after the compensation by D-STATCOM. The symmetrical and unsymmetrical fault study and performance is done by the given D-STATCOM Simulink model.

2.2 Controlling Scheme of D-STATCOM

A controller is utilized to sustain a continuous voltage magnitude at the value of point where a sensitive load is associated under system disturbances [11]. The VSC switching stratagem is based on a sinusoidal pulse width modulation (SPWM). This

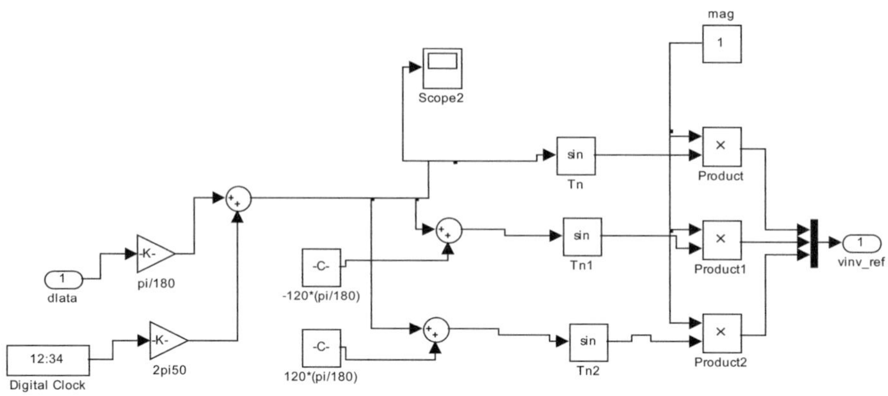

Fig. 2 Mathematical model of pulse width modulation (PWM) generator

technique compromises a more efficient flexible and reliable preference than the fundamental frequency switching (FFS) techniques preferred within FACTS applications [8]. High switching frequencies (HSF) are utilized to improve the effectiveness of the voltage converter without acquiring substantial switching losses [12]. The error signal is the input for controller which is acquired from the reference value and rms value of the voltage measured at the terminal point [16]. This error is processed through a PI controller that output is the angle ϕ which is delivered to the PWM signal generator [13]. In case of indirectly controlled converter, power exchanges with the network simultaneously of the active and reactive power. Error signal is achieved through relating the reference voltage within rms measured value near the load point [15]. The PI controller develops the error signal and produces requisite and suitable angle to determine the error to zero; i.e., the load rms voltage is taken back to the reference voltage [14] (Fig. 2).

The sinusoidal signal V_{Control} is phase modulated by means of the angle

$$V_{\text{A}} = \sin \omega t + \phi \tag{1}$$

$$V_{\text{B}} = \sin \omega t + \phi - \frac{2\pi}{3} \tag{2}$$

$$V_{\text{C}} = \sin \omega t + \phi + \frac{2\pi}{3} \tag{3}$$

The foremost constraints of SPWM technique are modulation of amplitude indeed of signal and (FMI) frequency modulation index (M_{f}) of the triangular signal. The amplitude index (AI) is reserved stable at 1 per unit in directive to achieve the component of maximum fundamental voltage at controller output. The switching frequency (F_{S}) is set at the point 1080 Hz. The frequency modulation index is specified by

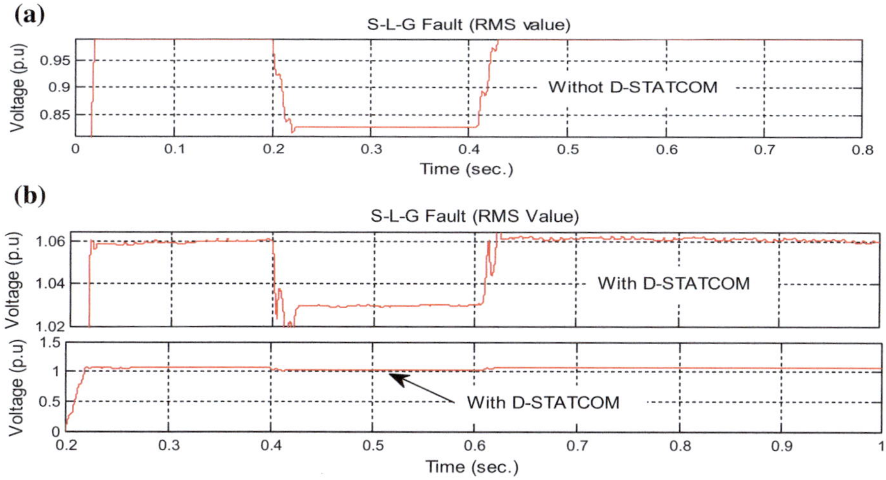

Fig. 3 **a** SLG fault waveform without D-STATCOM. **b** SLG fault waveform without D-STATCOM

$$M_{\mathrm{f}} = \frac{F_{\mathrm{S}}}{F_{\mathrm{f}}} = 1080/50$$

$$M_{\mathrm{f}} = 21.6 \tag{4}$$

where M_{f} is modulation index, F_{S} is switching frequency, and F_{f} is fundamental frequency

3 Performance of D-STATCOM Simulink Model Under Different Fault Conditions

3.1 Single Line-to-Ground Fault (SLG Fault)

In case of SLG fault, phase A is faulty and connected to the ground and phase B and phase C are unfaulty phases. The fault has impedance which is known as fault impedance denoted by Z_{F}. Phase B and phase C are like an open circuit, and there is no current flow through these phases.

In Fig. 3a, single line-to-ground (SLG) fault shows the result when D-STATCOM is not connected along with transmission line, and the rms voltage value is 0.80 per unit duration of 0.2–0.4 s.

Figure 3b shows the rms value is 0.99 per unit when D-STATCOM is coupled within transmission line and regulates the voltage near about to 1 per unit for time period of fault and attains the sinusoidal waveform at the load terminal in power distribution system.

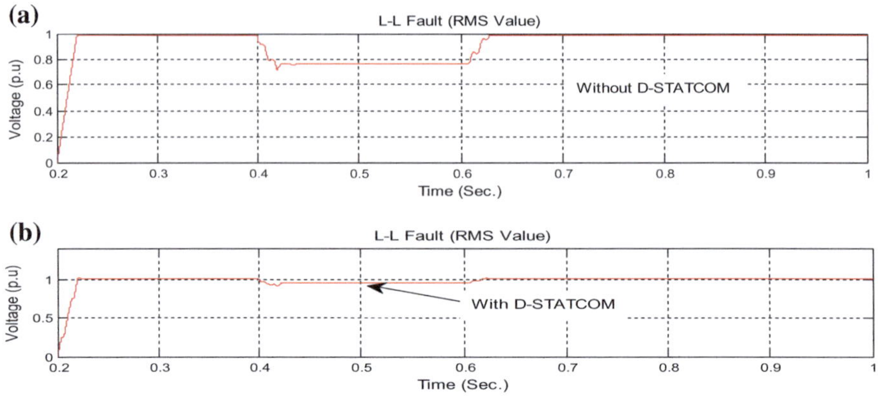

Fig. 4 **a** LL fault waveform without D-STATCOM. **b** LL fault waveform with D-STATCOM

3.2 *Line-to-Line Fault (LL Fault)*

In case of LL fault, fault takes place between two conductors, phase B and phase C, and both are short circuit; current flows through both phases. Fault impedance is Z_F.

Figure 4a presents the result without connected D-STATCOM along with distribution line, and rms voltage value is 0.75 per unit which is less than 1 per unit, and the sinusoidal waveform of system is not perfectly sinusoidal.

Figure 4b shows the voltage rms value is 0.95 per unit when D-STATCOM is utilized along with distribution line and regulates the voltage near to 1 per unit for duration of fault and achieves the sinusoidal waveform at the load terminal.

3.3 *Double Line-to-Ground Fault (LLG Fault)*

In case of LLG fault, phase B and phase C are short circuit and have their fault impedance value ZF, and both conductors are connected to the ground after a fault through a common line and have the value of ground impedance Z_g.

Figure 5a presents the result when D-STATCOM is not used in distribution system and shows the rms voltage value is 0.70 per unit and waveform is not sinusoidal.

Figure 5b shows the rms value is 0.93 per unit when D-STATCOM is utilized along with system. D-STATCOM regulates the voltage 0.70–0.95 per unit Regulated voltage is 0.25 per unit duration of fault, and achieved waveform is sinusoidal at the load terminal.

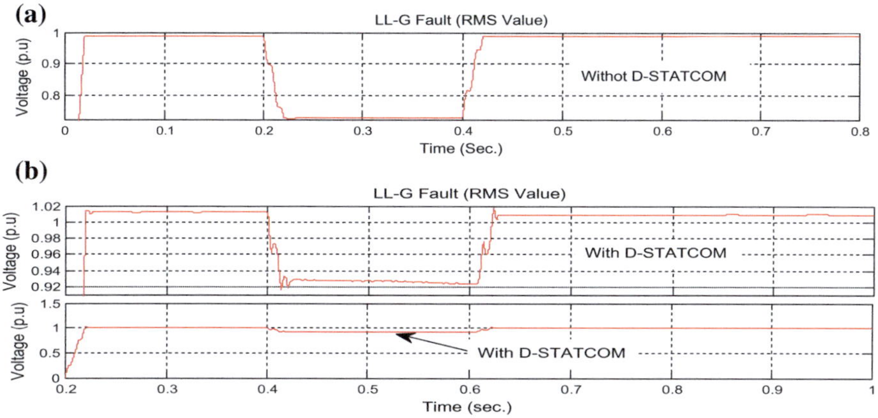

Fig. 5 **a** LLG fault waveform without D-STATCOM. **b** LLG fault waveform with D-STATCOM

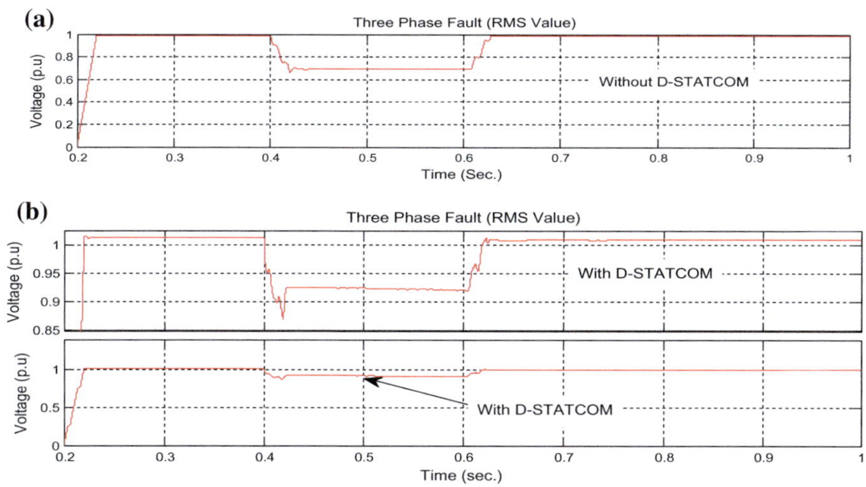

Fig. 6 **a** Three phase fault without D-STATCOM. **b** Three phase fault with D-STATCOM

3.4 Three Phase Fault

Three phase fault is known as symmetrical fault which involves all the three phases A, B, and C.

Figure 6a presents the rms voltage value that is 0.70 per unit duration of 0.4–0.6 s; this result is achieved when D-STATCOM is not connected along with distribution line. In this case, voltage sag takes place.

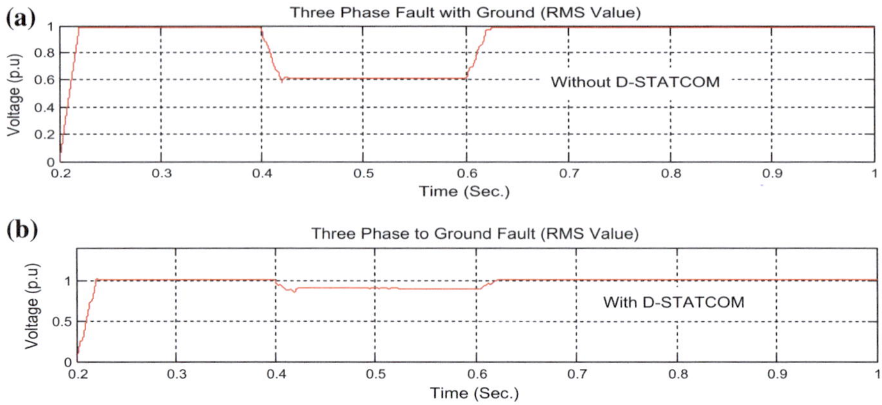

Fig. 7 **a** Three phase to ground fault without D-STATCOM. **b** Three phase to ground fault with D-STATCOM

Figure 6b shows the rms value is 0.94 per unit when D-STATCOM is connected with system. It regulates the voltage 0.7–0.94 per unit during fault occurring time (0.4–0.6 s). D-STATCOM is used to get sinusoidal balanced waveform at the load terminal in power distribution system.

3.5 *Three Phase-to-Ground Fault*

In three phase-to-ground fault, all the phases are involved with fault and connected with ground. This fault destroys the sinusoidal wave and causes to the unbalanced condition at the near nonlinear load.

Figure 7a presents the three phase-to-ground fault, and the result shows the rms voltage value is 0.60 per unit during the time 0.4–0.6 s; this duration is fault occurring time, and D-STATCOM is not linked within distribution line; the waveform is not sinusoidal at the load terminal.

Figure 7b shows the rms value is 0.90 per unit when D-STATCOM is attached along with system; it regulates the voltage 0.60–0.90 per unit D-STATCOM regulate value is 0.30 per unit during fault condition, and achieved waveform is sinusoidal at the load terminal in power distribution system.

4 Results and Discussion

Fault condition	Without D-STATCOM voltage (RMS value) (per unit)	With D-STATCOM voltage (RMS value) (per unit)	Power quality improvement (per unit)
SLG fault	0.80	0.99	0.19
LL fault	0.75	0.95	0.20
LLG fault	0.70	0.93	0.23
Three phase fault	0.70	0.94	0.24
Three phase-to-ground fault	0.60	0.90	0.30

5 Conclusion

The PWM voltage control scheme of D-STATCOM is utilized to regulate the load voltage under the symmetrical and unsymmetrical fault, and the D-STATCOM can regulate the load voltage to the desired level which is described in this paper. The performance of D-STATCOM has been evaluated for linear loads and static nonlinear loads. In this paper, the problems associated with faults determination and proper D-STATCOM injected reactive power rating which causes improvement in profile of voltage, current and reduction of power-losses in electrical power system network has been presented. Pulse width modulation control scheme is appropriate for both unbalanced load compensation and balanced RC or RL load. In this paper, the exploration of the D-STATCOM is carried out to improve the power quality in power distribution networks within static linear and nonlinear loads. Proportional integral (PI) controller is utilized within device to improve system performance. The results show the satisfactory performance of D-STATCOM in the distribution system under different fault conditions, and it can be concluded that D-STATCOM effectively improves the power quality and voltage profile in power distribution system.

Appendix

The constraints of the system considered in table:

Constraints	Parameters
Three phase source	20 kV
Three phase transformer	Nominal power = 100 MVA, frequency = 50 Hz
PI controller	$K_p = 0.5$, $K_i = 500$
Converter	3Arm-bridge, 6 pulse, carrier frequency = 50 Hz.
Load	RL load (non-linear load)
Switching frequency	1080 Hz

References

1. Kumar, A., Pal, N.S., Ansari, M.A.: Mitigation voltage sag/swell and harmonics using self-supported DVR. In Ist IEEE International Conference on Power Electronics, Intelligent Control and Energy Systems (ICPEICES-2016), pp 393–399 (2016)
2. Woo, S.M., Kang, D.W., Lee, W.C., Hyum, D.S.: The distribution STATCOM for reducing the effect of voltage sag and swell, IECON'01, vol. 2, pp. 1132–1137 (2001)
3. Park, S.Y., Park, J.K.: The modeling and analysis of shunt type custom power device. IEEE Power Eng. Soc. Winter Meet. **1**, 186–191 (2001)
4. Chiang, H.K., Lin, B.R., Yang, K.T., Yang, C.C.: Analysis and implementation of a NPC-based DSTATCOM under the abnormal voltage condition. In International Conference on Industrial Technology 2005, pp. 665–670, Dec 2005
5. Muni, B., Rao, S., Vithal, J.: SVPWM switched DSTATCOM for power factor and voltage sag compensation. In 2006 International Conference on Power Electronics, Drives and Energy Systems, PEDES'06, pp. 1–6, New Delhi, India (2006)
6. Kumar, C., Mishra, M.K.: A voltage-controlled DSTATCOM for power-quality improvement. IEEE Trans. Power Delivery **29**(3), 1499–1507 (2014)
7. Geddada, N., Karanki, S.B., Mishra, M.K.: Synchronous reference frame based current controller with SPWM switching strategy for DSTATCOM Applications. In Proceedings of IEEE International Conference on Power Electronics, Drives and Energy Systems (PEDES). Bengaluru, India, 16–19 Dec 2012
8. Singh, B., Jayaprakash, P., Dp, Kothari: A T-connected transformer and three-leg VSC based DSTATCOM for power quality improvement. Proc. IEEE Trans. Power Electron. **23**, 2710–2718 (2008)
9. Bilgin, H.F., Ermis, M.: Design and implementation of a current source converter for use in industry applications of D-STATCOM. Proc. IEEE Trans. Power Electron. **25**, 1943–1957 (2010)
10. Deben Singh, M., Khumanleima, Chanu L.: Power electronics technology for power quality improvement. Int. J. Adv. Res. Electr. Electron. Instrum. Eng. **4**, 2073–2080 (2015). https://doi.org/10.15662/ijareeie.2015.0404031
11. Ghosh, A., Jindal, A.K., Joshi, A.: Inverter control using output feedback for power compensating devices. In Conference on Convergent Technologies for Asia Pacific Region, pp. 48–52 (2003)

12. Singh, B., Jayaprakash, P., Kothari, D.P.: Three-phase four-wire DSTATCOM with H-bridge VSC and star/delta transformer for power quality improvement. In Proceedings of INDICON 2008, vol. 2, pp. 412–417, Dec 2008
13. Singh, B., Jayaprakash, P., Kothari, D.P.: Three-phase 4-wire DSTATCOM based on H-Bridge VSC with a star/hexagon transformer for power quality improvement. In: Proceedings of ICIIS 2008, pp. 1–6, Dec 2008
14. Tavakoli Bina, M., Eskandari, M.D., Panahlou, M.: Design and installation of a ± 250 kVAr D-STATCOM for a distribution substation, Elsevier, Electric Power Systems Research, vol. 73, pp. 383–391, 2005. https://doi.org/10.1016/j.epsr.2004.09.005
15. Tavakoli Bina, M., Eskandari, M.D.: Consequence of unbalance supplying condition on a distribution static compensator. In: IEEE Power Electronics Specialists (PESC'04), vol. 8, pp. 3900–3904, June 2004
16. Singh, B., Adya, A., Mittal, A.P., Gupta, J.R.P.: Power quality enhancement with DSTATCOM for small isolated alternator feeding distribution system. In: Conference on Power Electronics and Drive Systems, vol. 1, pp. 274–27 (2005)

Optimal Sitting and Sizing of Capacitor Using Iterative Search Method for Enhancement of Reliability of Distribution System

Vikas Singh Bhadoria, Nidhi Singh Pal, Vivek Shrivastava and Shiva Pujan Jaiswal

Abstract A secure and reliable distribution network is unavoidable prerequisite for socioeconomical development of any nation. Capacitors can be used for improving the voltage profile and reliability, and these can reduce reactive component of current. The aim of this paper is to present a mathematical model to fix an optimum size capacitor at optimum location in a radial distribution system for reactive power compensation. An iterative search technique is developed for loss minimization and voltage improvement of system. Authors also provide the effect of capacitor on the reliability of 34 radial bus systems.

Keywords Distribution system · Capacitor · Iterative technique · SAIDI · SAIFI

1 Introduction

The main objective of power system is to generate power and distribute among the different utilization points with high quality, reliability, and safety at passable economy. In deregulated power system, generation, transmission, and distributions are responsible for economical and reliable operation of power system. Over the past, utilities are less bothered about the reliability of distribution system, because of low economical benefit as compared to generation and transmission. Therefore, 90% of

V. S. Bhadoria · N. S. Pal
Department of Electrical Engineering, Gautam Buddha University, Greater Noida, India
e-mail: vikasbhadoria@gmail.com

N. S. Pal
e-mail: nidhi@gbu.ac.in

V. Shrivastava
Department of Electrical Engineering UEC, RTU, Kota, India
e-mail: shvivek@gmail.com

S. P. Jaiswal (✉)
Department of Electrical Engineering and Electronics Engineering,
Sharda University, Greater Noida, UP, India
e-mail: shivajaiswal@gmail.com

S. N. Singh et al. (eds.), *Advances in Energy and Power Systems*, Lecture Notes in Electrical Engineering 508, https://doi.org/10.1007/978-981-13-0662-4_11

customer's services are affected due to failure in distribution system [1]. Customers are compelled to face permanent load shedding, momentarily load shedding, or voltage sag due to the fault. But competitive market of power forced utilities to satisfy customer requirements by enhancing system adequacy and security. To minimize gap between demand and supply of electricity, it is necessary for power industry to integrate some equipment for loss minimization, reactive power compensation, power quality improvement, enhancement of voltage profile, and cost reduction with high level of customer's satisfaction. Distribution systems have high I^2R (up to 13%) as compared to transmission line because of loss, low voltage, and high current. The main problems of power utilities are high difference between peak load and valley load, perverse network structure, lack of reactive power compensation, and poor voltage regulation [2]. Reactive component of current in distribution network causes energy loss, voltage drop, and low power transfer capability of lines. In literature, numbers of techniques are available to counter the above problems such as distributed generation [3], Flexible AC Transmission System (FACTS) devices [4], network reconfiguration, load balance, On Load Tap Changer (OLTC), generator power factor control, power curtailment, energy storage, active filters, shun reactor and capacitor bank [5]. Losses in distribution systems, which are about 33.7–64.9% of the whole power system loss and poor power quality, are major problems. Higher losses result in low transfer capability of line by limiting the thermal limits and voltage limits. To improve the reliability and minimize the energy loss, capacitor bank is one of the best solutions for utilities. Capacitors are capable to reduce energy loss, voltage correction with low investment as compared to other technique. Capacitors are appearing to be a simple device which consist of two metal sheets separated by a dielectric insulating material with no moving part. Actually, the power capacitor is a complex device and technically developed with sophisticated manufacturing technique. Power capacitors use very thin dielectric materials which are capable to handle very high electric stresses. Due to the tremendous improvement in dielectric material, capacitor size increases from 15–25 KVAR to 200–300 KVAR range. Power capacitor manufacturers use all-film technology, which is more efficient than previous Kraft paper base capacitors [6]. Inappropriate selection of size and location of capacitor lead to uneconomical system and even imperil protection of distribution system. Various capacitor fixing methodologies are available in the literature such as analytical method which use mathematical model to find the suitable location and size of capacitor, numerical method based on iterative methods, genetic-based technique, heuristic method, artificial intelligence, and fuzzy logic [7]. The aim of this paper is to find the suitable size and location of capacitor using iterative technique [8]. Objective function is to minimize the total line loss and voltage profile improvement by reactive power optimization. Authors firstly minimize energy losses, and then, it is shown that reliability indices of system are also improved.

2 Problem Formulation

A set of nonlinear, algebraic power flow equations are solved by Newton–Raphson algorithms to find losses and voltage at each node;

$$S_i = P_i + jQ_i = V_i \sum_{k=1}^{n} Y_{ik}^{*} V_k^{*} \quad (1)$$

$$S_i = \sum_{k=1}^{n} |V_i||V_k| \mathrm{e}^{j\theta ik} (G_{ik} - jB_{ik}) \quad (2)$$

$$S_i = \sum_{k=1}^{n} |V_i||V_k| (\cos\theta_{ik} + j\sin\theta_{ik})(G_{ik} - jB_{ik}) \quad (3)$$

Resolving into active and reactive component,

$$P_i = P_{Gi} - P_{Di} = \sum_{k=1}^{n} |V_i||V_k| (G_{ik}\cos\theta_{ik} + B_{ik}\sin\theta_{ik}) \quad (4)$$

$$Q_i = Q_{Gi} - Q_{Di} = \sum_{k=1}^{n} |V_i||V_k| (G_{ik}\cos\theta_{ik} - B_{ik}\sin\theta_{ik}) \quad (5)$$

The active and reactive loss in a line between the busses can be calculated by

$$P_{\text{loss}}(i.k) = \frac{P_m^2 + Q_m^2}{|V_i - V_k|^2} \times R_m \quad (6)$$

$$Q_{\text{loss}}(i.k) = \frac{P_m^2 + Q_m^2}{|V_i - V_k|^2} \times X_m \quad (7)$$

From Eqs. (6) and (7), active and reactive loss of the system is equal to the addition of all line power loss and given by

$$P_{\text{T,loss}} = \sum_{k=1}^{n} P_{\text{loss}}(i, k) \quad (8)$$

$$Q_{\text{T,loss}} = \sum_{k=1}^{n} Q_{\text{loss}}(i, k) \quad (9)$$

Total apparent loss is given by Eq. (10)

$$S_{\text{T,loss}} = \sqrt{P_{\text{T,loss}}^2 + Q_{\text{T,loss}}^2} \quad (10)$$

Objective Function

In this paper, iterative method is used to find the optimal location and optimum size of capacitor and objective is to minimize total losses. The mathematical expression of objective function is given by

$$\text{Minimize}(f) = \min\left\{S_{\text{T,loss}}\right\} \tag{11}$$

Reliability Indices

The proposed optimization technique is not only to improve performance of distribution system but also improve reliability indices such as System Average Interruption Frequency Index (SAIFI) and System Average Interruption Duration Index (SAIDI). In a predefined time period, how many times customers are subjected to average number of interruption is indicated by SAIFI. SAIDI indicates the total duration of interruption on an average customer in given time interval [9]. Mathematically, SAIFI and SAIDI are given by Eqs. (12) and (13)

$$\begin{aligned}\text{SAIFI} &= \frac{\text{Total number of customer interruption}}{\text{Total number of customer served}} \\ &= \frac{\sum_{i=1}^{n} \lambda_i N_i}{\sum_{i=1}^{n} N_i} \text{interruptions/system customer}\end{aligned} \tag{12}$$

SAIDI indicates that the total duration of interruption on an average customer in given time interval is

$$\begin{aligned}\text{SAIDI} &= \frac{\text{Sum of customer interruption duration}}{\text{Total number of customer served}} \\ &= \frac{\sum_{i=1}^{n} U_i N_i}{\sum_{i=1}^{n} N_i} \text{hrs/system customer}\end{aligned} \tag{13}$$

where λ = failure rate, N = number of customer, U = annual outage time, i = load point.

3 Result and Analysis

In this paper, a 34-bus radial distribution system is taken as a test case. Line data and bus data as well as data required to calculate reliability indices, such as numbers of customer, failure rate, and repair rate of each component, etc., are taken from [10]. Table 1 demonstrates the results of the iterative technique. From results, it is clear that installation of 3.3 MVAR capacitor with failure rare 0.1156 f/year and repair rate 3 h at bus number 8 will result in 23% loss reduction and significance improvement of reliability indices.

Table 1 Results of the iterative technique

Case	Power loss (kW)	SAIDI	SAIFI
Without capacitor	221.72	6.088135	1.27333
With capacitor	170.06	6.06275	1.26068

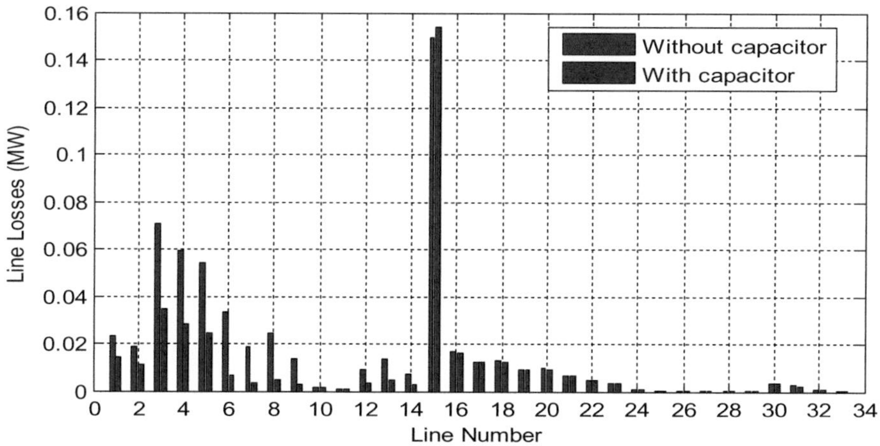

Fig. 1 Active power losses in each line with and without capacitor

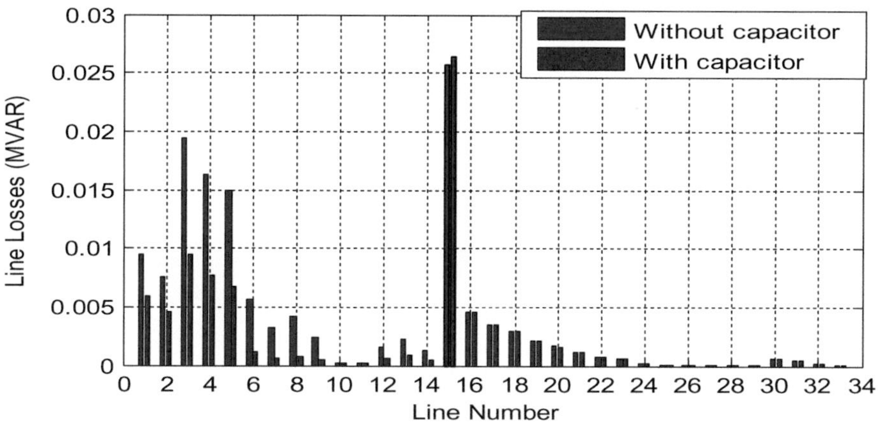

Fig. 2 Reactive power losses in each line with and without capacitor

Figures 1 and 2 show that active and reactive power loss in each line is significantly reduced by addition of capacitor; i.e., transfer capability of lines is also enhanced.

Installation of capacitor will also improve the voltage profile of the system. Voltage profile of the system with and without capacitor is shown in Fig. 3.

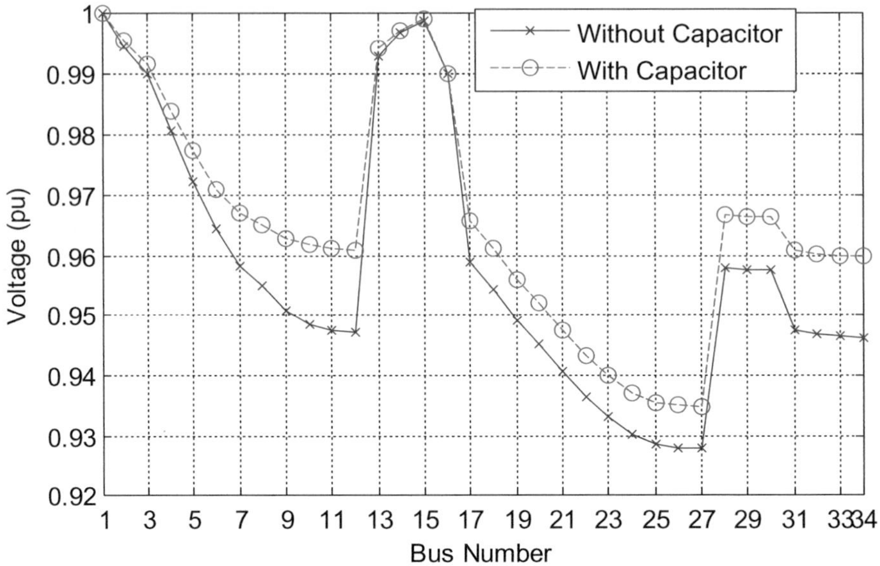

Fig. 3 Voltage profile of the system with and without capacitor

4 Conclusion

An iterative technique is used for the size and location optimization of capacitor in 34-bus distribution system. Results show the effectiveness of the algorithm. Improvement in SAIDI and SAIFI shows that optimal installation of capacitor also improves the overall reliability of the system. Although voltage parameter is not included in the objective function during the optimization process, still this installation enhances the voltage profile of the system. Overall, it can be concluded that optimum installation of capacitor not only reduces the system losses but it enhances the reliability and voltage profile of the system.

References

1. Esmaeilian, H., Darijany, O., Mohammadian, M.: Optimal placement and sizing of DGs and capacitors simultaneously in radial distribution networks based on voltage stability security margin. Turk. J. Elect. Eng. Comp. Sci. 1–14 (2014)
2. Mohammad, A., Masoum, S., Ladjevardi, M., Jafarian, A., Fuchs, E.F.: Optimal placement, replacement and sizing of capacitor banks in distorted distribution networks by genetic algorithms. IEEE Trans. Power Del. **19**(4), 1794–1801 (2004)
3. Bhadoria, V.S., Pal, N.S., Shrivastava, V.: Installation of DG for Optimal Demand Compensation, ICICT, pp. 821–824 (2014)

4. Jaiswal, S.P., Shrivastava, V.: Allocation of UPFC in distribution systems to minimize the losses. In Computational Intelligence on Power, Energy and Controls With Their Impact on Humanity (CIPECH), IEEE, pp. 202–205 (2014)
5. Chung-Fu, C.: Reconfiguration and capacitor placement for loss reduction of distribution systems by ant colony search algorithm. IEEE Trans. Power Sys. **23**(4), 1747–1755 (2008)
6. Gonen, Y.: Application of capacitors to distribution systems (Chap. 8). In: Electric Power Distribution System Engineering 2nd edn, pp. 371–493, CRC press (2007)
7. Da Silva, I.C., Carneiro, S., Sandoval, J., de Oliveira, E.J., de Souza, Costa J., Pereira, J.L.R., et al.: A heuristic constructive algorithm for capacitor placement on distribution systems. IEEE Trans. Power Sys. **23**(4), 1619–1626 (2008)
8. Bhadoria, V.S., Pal, N.S., Shrivastava, V.: Comparison of analytical and heuristic techniques for multi-objective optimization in power system (Chap. 13). In Saxena, P., Singh, D., Pant, M. (eds) Problem Solving and Uncertainty Modeling Through Optimization and Soft Computing Applications. Hershey, PA: IGI Global (2016). https://doi.org/10.4018/978-1-4666-9885-7
9. Billinton, R., Allan, R.N.: Distribution systems: basic techniques and radial networks (Chap. 7). In: Reliability Evaluation of Power Systems, 2nd edn, pp. 220–246. Plenum Press, New York, London (1996)
10. Kumar, D., Samantaray, S.R., Kamwa, I., Sahoo, N.C.: Reliability-constrained based optimal placement and sizing of multiple distributed generators in power distribution network using cat swarm optimization. Electr. Power Compon. Syst. **42**(2), 149–164 (2014). https://doi.org/10.1080/15325008.2013.853215

Battery Energy Storage Technology Integrated for Power System Reliability Improvement

Same Ram Ramavat, Shiva Pujan Jaiswal, Nitin Goel and Vivek Shrivastava

Abstract In this decade, application of solar and wind power in all countries of world increases. This has resulted in various challenges thrown to engineers to improve reliability and stability of power systems. To improve the reliability, stability and operating conditions of power system by incorporation of energy storage system is presented in this manuscript. This paper explains present and future status battery storage technology, the cost and profit scenario and its overall effects in improvements in reliability of power systems. An effort has been made to explain various challenges and solutions thereof, in present-day environment of power supply systems. Importance of battery energy storage system has also been discussed in power systems. A case study of improvement of reliability using energy storage system has been presented. In the end of this paper, improvement in battery technology with a perspective to use in power system and battery-operated vehicles has been discussed.

Keywords LOLE · EIR · EENS · LOEE

1 Introduction

The requirement of energy storage system for smooth and efficient operation of power system has been felt by power engineers and scientists over a long period of time all over the world. Pros and cons of various storage devices/systems are in

S. R. Ramavat (✉) · S. P. Jaiswal · N. Goel
Department of Electrical and Electronics Engineering, Sharda University,
Greater Noida, UP, India
e-mail: samay552007@rediffmail.com

S. P. Jaiswal
e-mail: shivajaiswal@gmail.com

N. Goel
e-mail: nitingoel_12@yahoo.co.in

V. Shrivastava
Department of Electrical Engineering UEC, RTU, Kota, India
e-mail: shvivek@gmail.com

S. N. Singh et al. (eds.), *Advances in Energy and Power Systems*, Lecture Notes in Electrical Engineering 508, https://doi.org/10.1007/978-981-13-0662-4_12

discussion among stakeholders. In this paper, various energy storage technologies, their applicability in improvement of power systems, deficiencies and limitations of these technologies are discussed. In earlier times, focus was on only on lead–acid battery [1]. Cost benefit, technology advancement, limitations of technology, economic model, specification of battery storage system and their application in modern power distribution system were analysed [2]. The use of battery technology for improvement in reliability, quality and stability of power supplied has been elaborated. Voltage depression and long-duration interruptions are major challenges in present-day power systems. Various other prevalent storage technologies are also discussed here. But it is observed and proved that battery storage technology is most appropriate in power systems all over the world and is being used widely. It is a considered opinion of authors that battery storage technology will play a vital role in improvement of power supply quality, particularly in renewable source of energy system. Authors have discussed other storage technology and their important role in quality power supply. Report mentioned at Ref. [3] (Tech Report: October 2013) gives insight to a complete catalogue of wind power technology, steam power technology, hydro power technology and energy storage technology. The report also gives a future perspective, security and stability of electric supply and its impact on surrounding environment, effects of climatic changes on them, economic viability for future planning of power supply systems. It has been experienced that lesson has not been learnt despite so many reports available on ground to improve existing technology of energy storage so far, and very little effort has been made. The reasons behind this reluctance in adopting new studies seem to be due to availability of conventional power in abundance matching demand of power supply. There is available flexibility in operation of the generating stations. The integration of costly energy storage technology can hardly be justified. In addition to higher cost, knowledge and experience of power engineers and skilled manpower is lacking in adoption, operation and future development of storage technology.

The recent development and quality supply requirement has motivated planners, engineers and manpower to relook these storage technology. Due to adverse impact on environment, large power plants cannot be established firstly, and secondly operation at below or above the capacity due to economic viability cannot be resorted to. The power availability from nonconventional sources is not steady and varied due to variable input resource like wind in some duration of day and season; sunlight during rainy and cloudy season. Most of nonconventional power plants are likely to supply as a distributed generation and stand-alone plants. In that case, these power plants can be made to operate as a local grid, and in that case, a strong case can be made to use energy storage technology to improve quality and quantity of supply to consumers [4].

The storage technologies available on ground are studied and evaluated, but battery energy storage technology is thoroughly investigated in the context of economic viability, likely future development and its importance in improvement of quality and quantity of electric supply. If a study of Danish electricity grid is made, then importance of battery storage in improvement of electric supply in terms of stability and reliability can be understood [5].

2 Technology Used in Energy Storage

There are two parts of the technology, i.e. battery itself, power conditioning parts and rest is control system. There is a rapid development in improvement of battery system and its power delivery system. In the following paragraph, each type of battery technology has been discussed [6, 7].

2.1. The batteries are made of group of cells which are made of plates of various conducting materials. The electrical energy is produced from chemical reaction, and when the batteries supply power, reverse reaction takes place. DC voltage generated by one cell can be connected in various combinations of series and parallel to deliver power at required voltage and current. Voltage and current ratings are as designed by designers. Efficiency and lifespan are also important criteria in design. The temperature at which batteries are operated and depth of discharge are also factored into design.

2.2. There are various types of batteries available in market which are commercially viable. Research and development is going on in some battery technology for development and testing and to make them commercially viable in time to come. Some of most suitable battery technology which is in use in power system is described below.

(a) *Lead–acid battery*: Two electrodes in battery are made of lead oxide and lead sponge. It is filled with hydro sulphuric acid so that all lead plates are completely submerged. All plates are contained in plastic cases suitable for heavy-duty application. The chemical reaction takes place where sulphuric acid reacts with lead oxide and lead sulphate is produced along with negative ions. The negative ions are collected on positive electrode, and thus, electricity is produced. During discharge, reverse action takes place.

(b) *Pressure valve-regulated-type battery*: Lead–acid battery is closed under pressure with a pressure-regulated valve. The movement of electrolyte is also restricted.

(c) *Sodium–sulphur battery*: In sodium–sulphur battery, positive and negative electrodes are made with molten sulphur and molten sodium, respectively. The electrolyte is solid beta alumina ceramic. It has a unique quality to allow passing positive sodium ions and a chemical reaction of combining sodium ion with sulphur to produce sodium polysulphide, and electric current flows in outer circuit generating 2 V. To facilitate this chemical process, a temperature of 300 °C is maintained surrounding it.

(d) *Lithium-ion battery*: Anode and cathode are made of graphite carbon in layers and lithium metal oxide, respectively. The electrolyte is made with a combination of lithium salt in organic carbonate. During charging process, lithium atoms on cathode electrode are converted into ions and pass through medium of electrolyte to anode and get deposited there after combining with electrons of external electric source. This process gets reversed during discharge or supply of power from this battery.

(e) *Metal–air battery*: These types of batteries are made of aluminium or zinc metal anode and porous carbon formation or metal mesh fully enveloped with catalyst. The electrolyte is normally sodium hydroxide.

(f) *Flow battery*: This type of storage technology is different from conventional battery bank. This consists of two tanks of anode and cathode duly separated by a membrane, and electrolyte is pumped by an external pump of suitable specification. Thus, chemical energy is converted into electrical energy when electrolyte passes through cathode and anode. Two electrolytes are kept in large storage tanks for circulation as per design requirement of electric energy density. The size of storage tanks and quantity of electrolyte flowing determine energy density of the battery storage system and rate of chemical reactions at anode and cathode. These batteries are also called redox flow batteries. These batteries are of high power rating, long duration and high energy rating. The self-discharge is zero as both electrolytes are stored separately outside. Flow batteries are of three types:
   (i) Regenerative fuel cell (polysulphide bromide)
   (ii) Vanadium redox (VRB)
   (iii) Zinc bromine (ZnBr)

2.3 *Power conditioning and control*: Power conditioning and control is a very important part of battery storage technology. The main function of the control and power conditioning is to regulate charging and discharging of battery and to establish a healthy interface with power supply load. It is maturing to establish its supremacy and utility in improvement of power supply to end-users in terms of reliability, stability and adequacy. Power solid-state electronic devices are used in a large scale in designing of control system for battery storage technology, and their latest development and availability has been harnessed. Major application is in most complicated interconnected and stand-alone power distribution system. In some countries, studies were carried out for improvement of power system in mini grids. The size of battery storage system is not recommended in this paper as it is driven by commercial considerations and profitability.

3 Battery Storage for Power System Reliability and Stability

Several studies have been made to understand the use of battery storage in improvement of power system operation and its impact on economy of power system. The application of battery storage on two sides, i.e. on utility side and on demand side. However, it is used for load levelling application and power quality [8]. The effort

Table 1 Generation data of system

Generation data		
Unit	Capacity (MW)	Probability
Unit 1	0	0.05
	15	0.3
	25	0.65
Unit 2	0	0.03
	30	0.97
Unit 3	0	0.04
	25	0.96

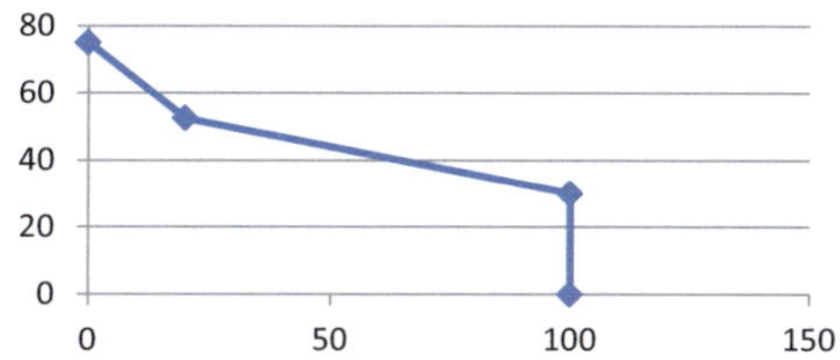

Fig. 1 Load duration curve

Table 2 Load data

Duration (hrs)	Power (MW)
000	75.0
020	52.5
100	30.0
100	00.0

was made to determine a method for assessing reliability of various generating units simultaneously in parallel to have a fair idea for planning.

Case Study: A power plant has three units of 25, 30 and 20 MW [9]. Probability of generating power has been given in Table 1. Maximum demand of power is 65 MW. Energy Index of Reliability (EIR), Loss of Load Expectation (LOLE), Expected Energy Not Supplied (EENS) and Loss of Expected Energy (LOEE) have been calculated with or without energy storage of capacity of 5 MW with probability 0.99 (Fig. 1; Tables 2, 3 and 4).

$$\text{EIR} = 1 - \frac{\text{EENS}}{\text{Total Energy Supplied}} \tag{1}$$

From above analysis, it is observed that there is improvement in all reliability indices when energy storage system is used, and comparison of indices is shown in Figs. 2, 3, 4 and 5; using bar chart, in each figure bar 1 represents without storage and bar 2 represents the indices with storage.

Table 3 Calculations of reliability indices without storage

Sl. No.	Capacity available (MW)	Probability	Cumulative probability	Energy curtailed	EENS
1	75	0.60528	1	0	0
2	65	0.27936	0.39472	44.4	12.403
3	55	0.02522	0.11536	177.8	4.484
4	50	0.04656	0.09014	286	13.316
5	45	0.03036	0.04358	475	14.421
6	35	0.00864	0.01322	1119.4	9.672
7	30	0.00194	0.00458	1575	3.055
8	25	0.00078	0.00264	2075	1.618
9	20	0.00144	0.00186	2575	3.708
10	15	0.00036	0.00042	3075	1.089
11	0	0.00006	0.00006	4575	0.279
				EENS	64.04
				EIR	0.986002
				LOLE	0.11536
				LOEE	0.01399

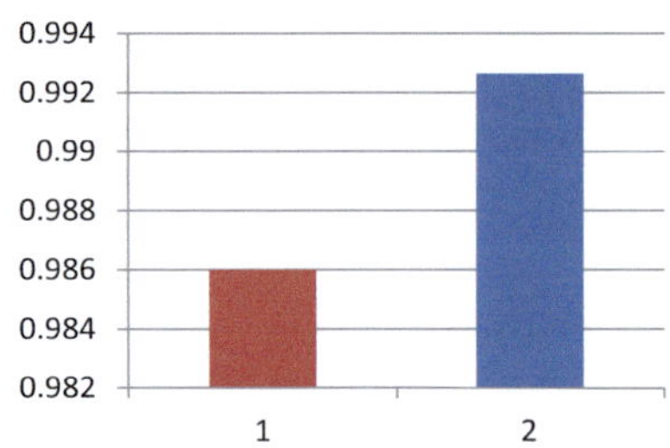

Fig. 2 EIR without and with storage

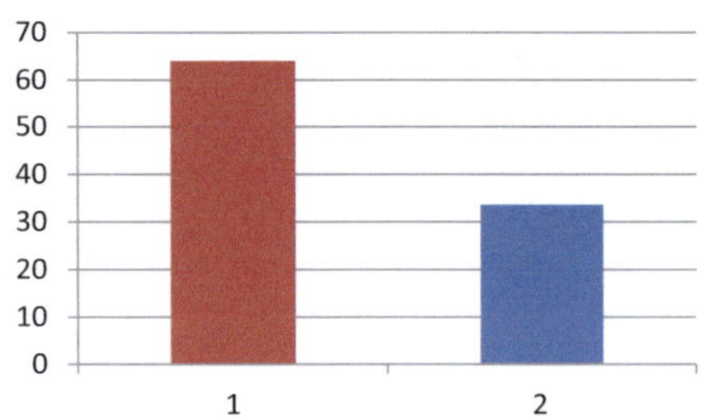

Fig. 3 EENS without and with storage

4 Future Applications of Storage Energy Systems

It is estimated that future of battery energy storage system will be bright and promising. The storage system can be used in the following circumstances [10]:

Table 4 Calculations of reliability indices with storage

Sl. No.	Capacity available (MW)	Probability	Cumulative probability	Energy curtailed (MWH)	EENS
1	80	0.59922	0.9879256	0	0
2	75	0.00605	0.3887056	0	0
3	70	0.27656	0.3826556	11.111	3.072
4	65	0.00279	0.1060956	44.4	0.123
5	60	0.02446	0.1033056	100	2.446
6	55	0.04634	0.0788456	177.8	8.239
7	50	0.01899	0.0325056	286	5.431
8	45	0.000303	0.0135156	475	0.144
9	40	0.00855	0.0132126	752.8	6.436
10	35	0.00201	0.0046626	1119.4	2.251
11	30	0.00079	0.0026526	1575	1.244
12	25	0.00143	0.0018626	2075	2.967
13	20	0.00037	0.0004326	2575	0.952
14	15	0.000003	0.0000626	3075	0.092
15	5	0.000059	0.0000596	4075	0.24
16	0	0.0000006	0.0000006	4575	0.002
				EENS	33.638
				EIR	0.992645
				LOLE	0.1033056
				LOEE	0.007352

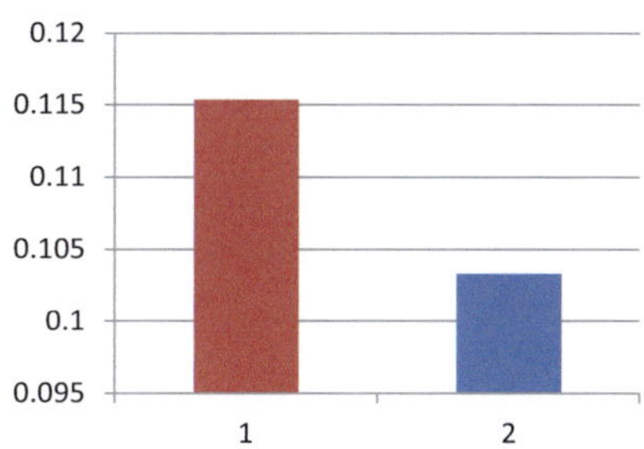

Fig. 4 LOLE with and without storage

(1) Instantaneous application: Large amount of power is required for a short duration of few seconds only for power stability, frequency control, spinning reserve, etc.
(2) Short-term application: Good amount of power is required for frequency regulation, power output levelling and harmonic compensation for few minutes.
(3) Midterm utility application: High amount of energy is required for peak load balancing, reliability improvement of renewable energy system, isolated grids and micro-grids.

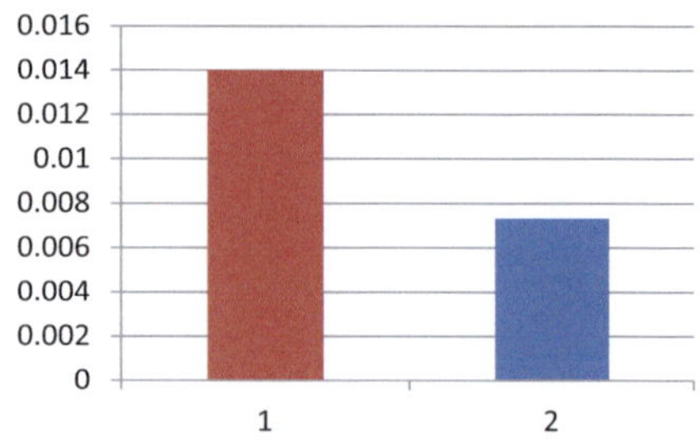

Fig. 5 LOEE without and with storage

(4) Long-term applications: Very amount of power is required so as to defer setting of new plant/renovation of capacity due to economic factors to supply peak load for large duration.

First three duration supply through battery storage system is economically viable but supply for fairly large duration may not be economical at all, i.e. more than 5 h. This is due to quick and instantaneous response of system as per changing requirements of power supply where in such situation synchronous generators cannot cope with. In future when solar power and wind power may be generated more than demand of electricity due to availability of many sunny days and high wind velocity during some part of the year, power can be stored in battery banks and stored power can be utilized during lean period. The cost of generation of renewable energy is decreasing day by day due to improvement and availability of new technology and becoming competitive with hydro and thermal generation.

Smoothening of power purchase prices:

In present system of purchasing bulk power from generating agencies or power trading company, an estimated quantity of power required by distributors are projected and bid one day advance. The estimate of power bid and actual demand may have large variations, thus unabridged gap between demand and supply of power. Spot purchase is very costly and may not be even available with power trading company. At this juncture, storage battery power will come to rescue situation of distributors, thus ensuring reliability of power supply to consumers [11].

By suitable placement of battery storage bank at load centres, power quality can be enhanced. Regional grids are interconnected through national grid for transmission of power from power surplus regions to power-deficient regions. During outage of national grids, regional grids work in isolated/island mode. The battery storage system is boon in ensuring regular supply to consumers. A situation of grid failure and resulting tripping of generating units may occur; huge power is required for restarting of units for a short duration. An ideal remedy of situation is battery storage energy. Indian government has projected 175 GW of solar and wind power by 2020 [12]. Enough transmission capacity will not be available to take this power to load centres. In such cases also, battery storage energy will be a suitable remedy.

5 Future Development Battery Energy Storage Technology

The author has elaborated various battery technologies earlier in this paper; the flow batteries will find the applications in future systems of high power supply system as well as high reliability. Lead–acid batteries are cycled very quickly, and flow batteries are having long working life. Lead–acid batteries will continue to find application in future. NaS and NiCd batteries will find application in mid-duration time frame, and Li-ion batteries will be used in portable equipment and applications. Designers can design control and conditioning of battery energy storage system in such a manners that these can behave like synchronous generators [13]. The flexibility of using battery is already very high and suites the requirements. The system can also be designed to be suitable to use from few second to few hours. The investigation is recommended in power systems in the light of present improved technology and liberated environment.

6 Conclusion

There is a promising future of battery storage technology in modern power supply system for improvement of reliability and stability. The designers are to work on improvement of technology to bring down cost of production of batteries, design of control and conditioning to enhance flexibility of rugged items so that they are suitable for site atmospheric conditions. The development of tools and plants, training of manpower to enhance their skill for smooth maintenance and operation are major area of future considerations. A major factor will be to design high density bulk energy storage and reduced idle discharging rate. Thus reliability, stability and economically viability of power supply system integrating energy storage can be suitably enhanced. The consumers demand to supply quality power with high reliability index can be met with. Results of case study show that all reliability indices are improved by application of battery.

References

1. Joseph, A., Shahidehpour, M.: Battery energy storage systems in electric power systems, pp. 1–8. IEEE (2006)
2. Roberts, B., McDowall, J.: Commercial successes in power storage. IEEE Power Energ. Mag. **3**(2), 24–30 (2005)
3. Technology energy data for energy plants, Danish Energy Agency, Oct 2013
4. Tahllam, R.S., Eckroad, S.: Multimode battery energy storage for custom power applications. In: Proceedings of the IEEE Power Engineering Society Winter Meeting, vol. 2(31), pp. 1147–1150, 31 Jan–4 Feb 1999

5. Zhan, C., Barnes, M., Ramachandaramurthy, V., Jenkins, N.: Dynamic voltage restorer with battery energy storage for voltage dip mitigation. In: Proceedings of the Eighth International Conference on Power Electronics and Variable Speed Drives, pp. 360–365, 18–19 Sept 2000
6. Sandia national laboratory. www.sandia.gov/ess
7. Electricity storage association. www.electricity-storage.org
8. Zeng, J., Zhang, B., Mao, C., Wang, Y.: Use of battery energy storage system to improve the power quality and stability of wind farms. In: International Conference on Power System Technology, Power on 2006, 1–6 Oct 2006
9. Billion, R., Allan, R.N.: Reliability evaluation of power system, 2nd edn. Plenum, New York (1996)
10. Strategic Energy Analysis and Applications Center: NREL National Renewable Energy Laboratory (US) (2009)
11. Shah, V.A. et al.: 15th National Power Systems Conference, Indian Institute of Technology-(IIT) Bombay (2008)
12. Strbac, G., Aunedi, M., Pudjianto, D., et al.: Strategic assessment of the role and value of energy storage systems in the UK low carbon energy future. Report for Carbon Trust. Imperial College London. Published June 2012. http://www.carbontrust.com/media/129310/energy-storage-systems-role-value-strategic-assessment.pdf. Accessed 26 May 14
13. Taylor, P., Bolton, R., Stone, D., Zhang, X.P., Martin, C., Upham, P., et al.: Pathways for energy storage in the UK. Technical report. Centre for low carbon futures. Published 27th March 2012. http://www.lowcarbonfutures.org/reports/research-reports?page=1. Accessed 11 June 14

Thermal Evaluation and Oxidation Stability of High Temperature Alternative Solid Dielectrics of Power Transformers in Mixed Oil

Manoj Kumar, Chilaka Ranga, Ashwani Kumar Chandel and Ashish Kumar Mishra

Abstract In this paper, a combination of mineral oil and synthetic ester oil in a standard mass ratio is used as an alternative liquid dielectric for power transformers instead of traditional mineral oil. In these mixed oils, the two alternative solid dielectrics of transformers, namely thermally upgraded Kraft and Nomex-910, are impregnated. An accelerated aging test is performed on these samples at different temperatures ranging from 100 to 180 °C. Further, ultraviolet spectroscopy test is performed in order to determine the thermal degradation as well as oxidation rate of these samples. It has been observed from the test results that Nomex impregnated mixed oil samples have lower oxidation rate as compared to that of thermally upgraded mixed oil samples. Therefore, Nomex impregnated transformers have lower oxidation rate and longer service life as compared to TUK impregnated transformers. The present experimental study is very beneficial to the transformer manufacturing companies as well as utilities.

Keywords Transformer · Blend oil · Insulating paper · Aging
Ultraviolet–visible spectroscopy · Oxidation stability

M. Kumar (✉) · C. Ranga · A. K. Chandel · A. K. Mishra
Electrical Engineering Department, National Institute of Technology Hamirpur, Hamirpur, India
e-mail: manojeeemit@gmail.com

C. Ranga
e-mail: rangasun.chilaka@gmail.com

A. K. Chandel
e-mail: ashchandelin@gmail.com

A. K. Mishra
e-mail: srtforeverashish@gmail.com

S. N. Singh et al. (eds.), *Advances in Energy and Power Systems*, Lecture Notes in Electrical Engineering 508, https://doi.org/10.1007/978-981-13-0662-4_13

1 Introduction

Transformer is one of the most important equipment of electrical power transmission and distribution systems [1, 2]. Reliability and long operational life of power transformer are highly desirable. Both reliability and life of the transformer are determined by its health condition and degradation rate of its insulation. Degradation of insulation in a transformer caused due to the various stress present in the transformer such as electrical, mechanical, chemical, and thermal stress. These stresses mainly influence the degradation rate of the transformers. In general, mineral oil and thermally upgraded Kraft are used as tradition insulation inside the transformers. Mineral oil is a petroleum-based product which is used in power transformer for more than 100 years, and it has long successful history [1–3]. Despite its long-term success mineral oil is facing criticism because of its low fire point, flash point, low breakdown voltage, and higher degradation rate. Mineral oil is not biodegradable spillage of oil in the event of fault or leakage create major environmental hazard [3, 4]. Also with the increase in environmental concern, we are looking for eco-friendlier alternative. Moreover, limited stock of petroleum product will limit its use in future as insulating liquid in transformer [5].

During the past few decades, researchers across the world are working in this area in order to investigate an alternative liquid dielectric for the transformers [4, 5]. Some of the researchers proposed that the combination of mineral oil with synthetic ester oil offer better performance than that of mineral and synthetic ester oils separately. Also it is found that a blend of mineral and synthetic ester oils in 80 and 20% proportion is suitable for both technical and economical point of view [1–4]. With the advancement in technology and increase in population, the need of compact substation is ineludible. Thus, there is a great need for the equipment which occupies less space and high safety factor. Loading and size of the transformer are heavily dependent on the losses occurring in the transformer. To reduce the volume of power transformers and enhance their long-term operational reliability, it is desirable to develop a new kind of oil-paper insulation system which has higher dielectric strength, higher thermal conductivity and lower dielectric losses as compared to conventional insulation systems.

In this paper, an effort has been made to find the alternative solid and liquid dielectric which have better performance than traditional insulation, i.e., mineral oil and thermally upgraded Kraft. To find an alternative solid dielectric, a new paper is considered which is designed and manufactured by DuPont Company in 2014, i.e., Nomex which have higher thermal stability than thermally upgraded Kraft. Nomex have higher tensile and breakdown strength than thermally upgraded Kraft. Further, an accelerated thermal aging is performed at different temperatures. In order to analyze the insulation performance, UV-spectroscopy test has been performed. From the test results and visual examination, it is observed that Nomex have lower oxidation rate and higher temperature bearing capacity than thermally upgraded Kraft. It also has lower degradation coefficient and higher operational life.

2 UV–Visible Spectroscopy

UV–Vis spectroscopy (UV–Vis) deals with the interaction of matter with radiation of ultraviolet and visible region. Molecule absorbs energy when excite from lower energy state to higher energy state or release energy when it returns to lower energy state from higher energy state. This excitation between different energy level is known as transition. Transitions involve in UV–Vis are electronic transition.

UV–Vis is primarily governed by the Beer–Lambert law. In which Bear law and Lambert law deal with the concentration and path length of the absorber, respectively. Molecule absorbs light when it interacts with radiation; this absorbance gives the information about structure, composition, or concentration of the absorber. Because different molecule absorbs light at different wavelength, absorbance in the solution is directly proportional to the

(a) Concentration of the absorber and
(b) Path length of absorber through which light pass.

In UV–Vis path length of the cuvette is constant, henceforth absorbance only depends on the concentration or composition of the absorber. Sample molecules absorb energy when it explores to light, and that energy matches with the possible electronic transition within the molecule. Some of the light is absorbed as electron promoted to higher energy state. And this absorption is recorded by the spectrophotometer and record the wavelengths at which absorption occur. Because absorption depends on the different molecules present in the sample. So this spectroscopy is very helpful to find the degradation of transformer oil because as the oil aged it chemical composition changes also sludge formation due to degradation oil. That can easily be detected by this spectroscopy technique, and it plays an important role in diagnosis of the power transformer insulation.

There are two cuvettes, one is filled with testing sample and other is filled with heptane for reference purpose. With the help of half-mirrored equipment, the incoming beam splits into equal intensity beam, i.e., sample beam and reference beam which pass through testing sample and reference sample, respectively. Intensity of both beams is measured with spectrophotometer detector and compared. The absorption coefficient (A) which is defined as [6].

$$A = \log \frac{I_0}{I}$$

where I_0 is the intensity of the reference beam and I is the intensity of the sample beam. Here, I_0 remain same because reference cuvette taken here does not absorb light. So if intensity of sample cuvette also remain unabsorbed than $I = I_0$ and absorption $A = 0$ [6]. If sample beam absorbs light than absorbance coefficient change. From these, spectra are drawn between the absorbance values at different wavelength ranging from 360 to 600 nm [7].

Fig. 1 Samples placed in an air-circulated temperature oven

3 Experimental Work

3.1 Samples Preparation

Initially, the conical flasks were neatly cleaned with water and then dried out by PID controller based air-circulated temperature oven (500 ± 2 °C) which is set at 100 °C for 24 h. Later the flasks were cooled down to room temperature. Samples in the present work were prepared with both Nomex and TUK insulations, in a standard mass ratio as per IEEE Std C57.154-2012 [8] (Fig. 1).

Conical flasks are filled with a mix of 160 ml of mineral oil and 40 ml of synthetic ester oil. These flasks were kept in oven, and subjected to a temperature of 100 °C for 24 h so as to remove the moisture from the oil. Further, the Nomex (2 mill) and TUK (2 mill) paper put in two separate flask filled with mix which were put into oven at 100 °C. Sample also contains a copper strip (35 mm × 20 mm × 3 mm) to replicate the real operating condition of power transformer. Finally, all the samples were tightly sealed with rubber corks and covered by aluminum foil.

3.2 Aging Procedure

Initially, four samples of each TUK and Nomex were kept in temperature oven and thermally aged at 120 °C. After 24 h, one set of Nomex and TUK is removed from oven. Similarly, other sets of samples are removed from oven after 48, 72, and 96 h. After the completion of aging at 120 °C, eight samples are collected whereas four samples correspond to Nomex and the other four are having TUK, aged for 24, 48, 72, and 96 h. Same accelerated thermal aging procedure is performed for 150 and 180 °C temperatures. Finally, three sets each with 8 samples aged at 120, 150 and 180 °C.

4 Results and Discussion

When a sample is illuminated by monochromatic light, some light is absorbed by sample and some of it passes through the sample. As discussed in Sect. 2, the absorbed light mainly depends on the composition and concentration of the absorber. With thermal aging, oil degrades with passage of time. When UV–Vis test is performed on fresh samples then the amount of light absorbed is very less because of the absence of degrading product in oil. On the account of an increase in thermal stress on the samples, production of deterioration product takes place which gets dissolved in water and in turn changes the composition of oil. These dissolve content increase the absorption of light in the sample. Increase in the absorption of light indicates more degradation of transformer oil.

Figure 2 shows the variation in UV spectrums of different samples which are aged at 180 °C and at different time durations. It is noticed from the test results that the spectrum of each sample moves upward as the duration of thermal stress on the sample increase. It indicates the increase in absorbance value. Also it is clear that the absorbance value is more in TUK paper than that of Nomex paper.

Figure 3 shows the variation in absorbance coefficient as per time over accelerated temperatures. At lower temperatures, the absorbance value of TUK and Nomex is approximately same. Very less deviation in absorbance value indicates that the both papers degrade with similar rate at initial stage of heating. But, as the temperature increases the rate of increase in the absorbance value of TUK paper is more than that of Nomex paper. Higher rate of increase in absorbance value indicates higher degradation rate, it implies that the TUK paper degrades more rapidly.

Absorbance value is directly related to the oxidation stability of transformer oil [9]. Thus, the oxidation stability of Nomex paper is more as compared to TUK paper. Poor oxidation stability results in sludge [3] which increases the viscosity of transformer oil and hinders the flow of oil in cooling tube in transformer. Cooling is also one of the main functions of transformer oil. If it fails to provide cooling, then temperature of the transformer increases. Increase in the temperature has adverse effect on the insulation of the transformer. Also according to IEC 60076-7, the service life of insulation will be halved for every 6 °C increase in temperature. It is well known that the life of the transformer is directly related to the life of the insulation [10]. From this, we can conclude that Nomex paper insulation has more operational life. Thus, if we use Nomex paper in transformer than operational life of the transformer can be increased. Also from UV–Vis result, it is evident that Nomex paper has better thermal stability at higher temperature than TUK.

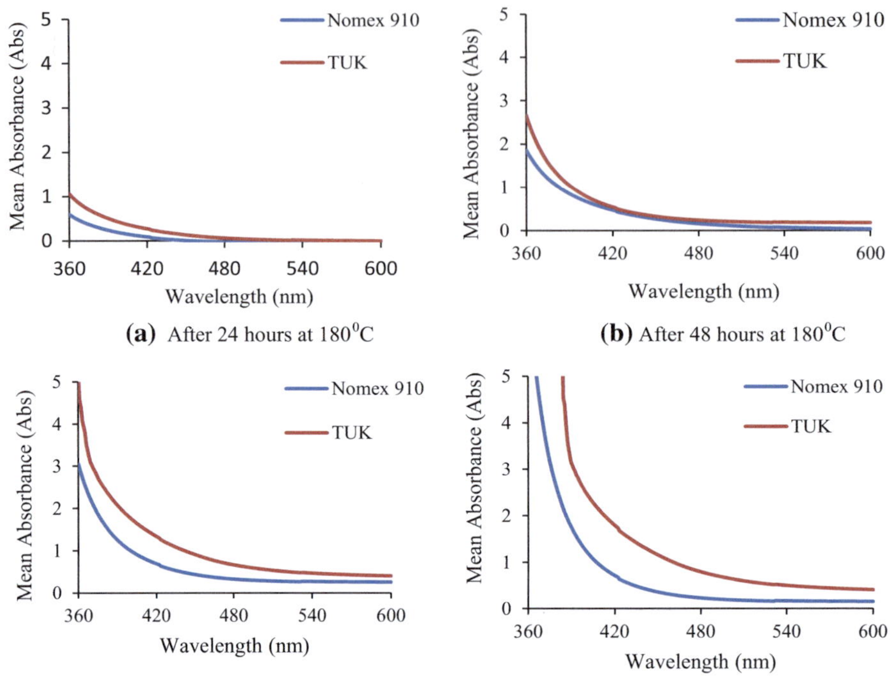

(a) After 24 hours at 180°C
(b) After 48 hours at 180°C
(c) After 72 hours at 180°C
(d) After 96 hours at 180°C

Fig. 2 UV–Vis absorption spectrums of TUK and Nomex impregnated mixed oil samples

4.1 Visual Observation

As oil is subjected to thermal stress the color of the oil becomes darker and its viscosity increases. Color of oil changes with increase in temperature and with the increase in duration of thermal aging. Change in color and physical appearance is critical observation to determine the condition of transformer from insulating oil. When sample is subjected under thermal degradation, both oil and paper get degrade. The degradation products from oil and paper will dissolve into the oil. Thus, diagnosis of oil gives indication about the overall degradation of transformer. From Fig. 4, it is evident that Nomex samples are less dark for same accelerated aging as compared to its TUK counterpart.

For more clear visual observation, paper is brought out from the flask. It is observed that the reduction in thickness in Nomex paper is less than that of TUK [9]. It was also observed that Nomex papers were physically stronger than TUK. This was evident from minute fragments which started to come out from TUK paper, when it was kept at higher thermal stress for more time period. Minute fragments settle down at bottom of the ampoules. This increases the darkness and increases the viscosity of the oil. In Actual practice with the increase in viscosity, fragment coming out

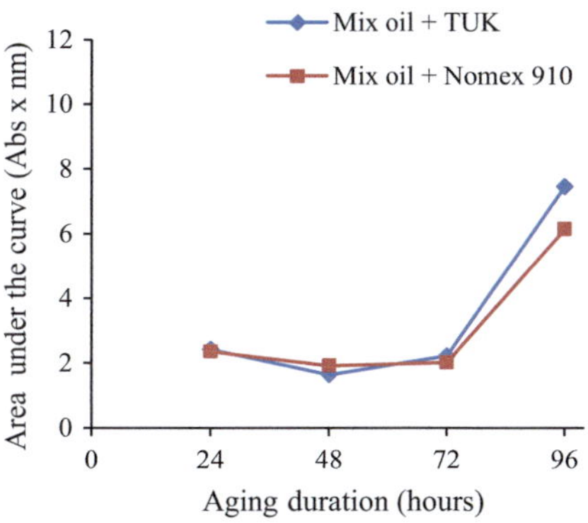

(a) At 180^{0}C temperature

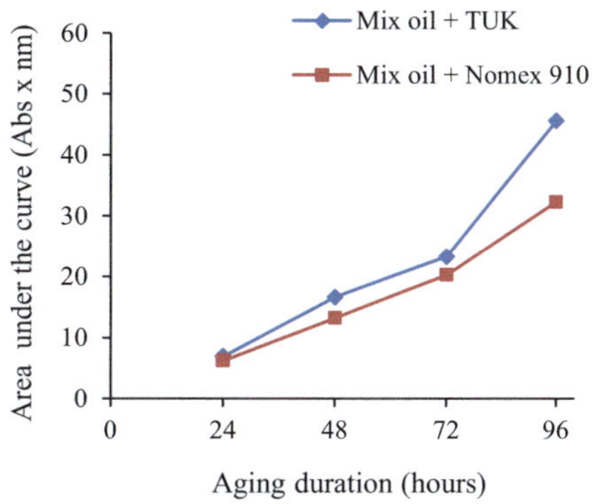

(b) At 180^{0}C temperature

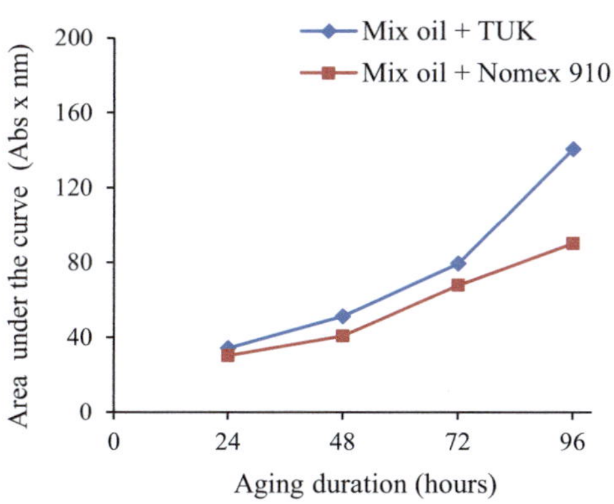

(c) At 180^{0}C temperature

Fig. 3 Variation in absorption value of TUK and Nomex-910 as a function of test duration

from paper and sludge formation hinder the flow of oil in cooling tube. Thus, the use of Nomex paper not only increases the oxidation stability but also improves the cooling performances of transformer oil. It has better thermal performance at higher temperatures.

5 Conclusion

In this paper, Nomex and traditionally used TUK paper are thermally aged at different temperature for different duration in mix oil in order to improve the insulation of power transformer. To analyze the effect of different temperature on degradation of Nomex and TUK paper UV–Vis test is performed. It is found that the absorb area under the spectra of TUK paper is more as compared to that of Nomex paper

Fig. 4 Color variation with aging temperature for TUK and Nomex-910 in mix oil after 72 h

absorbance area. Moreover, it is evident from UV–Vis results and visual observation that Nomex paper in combination with mixture of mineral and synthetic ester oils in 80 and 20% proportion is optimum from both technical and economical point of view. It offers better thermal and oxidation stability at higher temperatures. It also increases the operational life of the transformers.

Acknowledgements Authors are thankful to the authorities of TEQIP–II of NIT Hamirpur India for providing financial support with grant number NIT/HMR/TEQIP–II/Research and Development–9/2015/2157–63. Authors are also grateful to the authorities of TIFAC CORE Center of NIT Hamirpur India, and DuPont India for providing the necessary materials and facilities to perform the experiments of the present research.

References

1. Fofana, I., Wasserberg, V., Borsi, H., Gockenbach, E.: Challenge of mixed insulating liquids for use in high-voltage transformers, part 1: investigation of mixed liquids. IEEE Electr. Insul. Mag. **18**(3), 18–31 (2005)
2. Fofana, I., Wasserberg, V., Borsi, H., Gockenbach, E.: Challenge of mixed insulating liquids for use in high-voltage transformers, Part 2: investigation of mixed liquids Impregnated paper Insulation. IEEE Electr. Insul. Mag. **18**(4), 5–16 (2002)
3. Rao, U.M., Sood, Y.R., Jarial, R.K.: Oxidation stability enhancement of a blend of mineral and synthetic ester oils. IEEE Electr. Insul. Mag. **32**(2), 43–47 (2016)
4. Rao, U.M., Sood, Y.R., Jarial, R.K.: Performance analysis of liquid dielectric for power transformer. IEEE Trans. Dielectr. Electr. Insul. **23**, 2475–2484 (2016)
5. Perrier, C., Beroual, A., Bessede, J.L.: Improvement of power transformers by using mixtures of mineral oil with synthetic esters. IEEE Trans. Dielectr. Electr. Insul. **13**(3), 556–564 (2006)
6. Ranga, C., Chandel, A.K.: Oxidation stability of TUK and Nomex insulated power transformers. In IEEE Annual India Conference (INDICON) (2016)
7. ASTM E275-01 Standard.: Standard Practice for Describing and Measuring Performance of Ultraviolet, Visible, and Near-Infrared Spectrophotometers, vol. 03.06, pp. 72–81 (2001)

8. ASTM D923-07 Standard.: Standard Practices for Sampling Electrical Insulating Liquids (2007)
9. Zhang, X., Shen, S., Xu, Y., Lei, Q., Liu, G., Chen, R., Li, X.: Thermal evaluation of high-temperature insulation system for liquid-immersed transformer. In Electrical Insulation Conference (EIC) (2016)
10. Martin, D., Wijaya, J., Lelekakis, N., Sisa, D., Heyward, N.: Thermal analysis of two transformers filled with different oils. IEEE Electr. Insul. Mag. **30**(1), 39–45 (2014)

Multi-area Economic Dispatch Using Dynamically Controlled Particle Swarm Optimization

Vineet Kumar, Ram Naresh Sharma and Shiv Kumar Sikarwar

Abstract The goal of multi-area economic dispatch (MAED) is lower term resolution of optimal power generations of no. of electricity facilities to cater system demand at lowest feasible cost which can satisfy power balance, generating limit and transmission constraints. The generators watt and var power are controlled within the constraint of the generators' power limit to attain the minimum cost, satisfying the system burden. It is imperative to find an alternative route between two areas to supply spare power if the two areas are such that one of them is having surplus power as compared to other, or tie-line constraint joining the two areas is at transmission line. The power distribution of every unit is done in such that after supplying the total load, some specified reserve is left behind. In this paper, the comparison of classical PSO strategies and their variations for MAED has been accomplished. The ability to handle constraint of these meta-heuristic techniques allows them to produce high-quality solutions. The performance is subjected on a single area 3 generation units, a two area system with 4 generating units and a 4-area, 40-unit system with six tie lines.

Keywords Particle swarm optimization (PSO) · Particle swarm optimization time-varying acceleration coefficients (PSO-TVAC) · Dynamically controlled particle swarm optimization (DCPSO)

V. Kumar (✉) · R. N. Sharma · S. K. Sikarwar
Electrical Engineering Department, National Institute of Technology Hamirpur, Hamirpur, India
e-mail: vineetkmr192@gmail.com

R. N. Sharma
e-mail: rnareshnith@gmail.com

S. K. Sikarwar
e-mail: seeshivat@gmail.com

S. N. Singh et al. (eds.), *Advances in Energy and Power Systems*, Lecture Notes in Electrical Engineering 508, https://doi.org/10.1007/978-981-13-0662-4_14

1 Introduction

Economic dispatch problem is of considerable importance in the area of power generation. Economic dispatch is the determination of electrical power outputs from all the thermal generators supplying power to the system at minimum total cost in a single area subject to the condition that total meets the load demand including losses while power capacity limit constraints of generators are satisfied. Multi-area economic dispatch (MAED) is formulated by extending ED model. The goal of MAED is to find out electric power outputs of committed generators in all the interconnected areas at minimum cost while satisfying load balance constraints, electric power capacity limits of generators and transmission line capacity constraints. An electrical power system consists of generation, transmission and distribution utilities to enhance electrical power to the consumers. ED problem for single area power system is addressed by most of the research papers, whereas only few papers address ED for multi-area power system but the optimization problem still solved in a sequential way. The objective is to find the solution MAED problem using a developed Lagrange's decomposed method. Large networked systems are break down into areas or zones. MAED problem is a methodology in power network operation that allows measure of generation to committed units within the areas. Earlier ED problem is solved by lambda iteration method, gradient method, reduced gradient method, Newton method and other methods like participation factor method, binary weighted method. All these classical techniques have several disadvantages like: problem complexity, time computation factor, characteristics should be non-convex in nature. All these factors are removed by other classical methods like priority method, dynamic programming and Lagrange relaxation method. These methods have several advantages that these techniques are applied to complex characteristics problems of generator but there is problem in these techniques that they have take several stages to solve the problem and time taken is too large [1]. These things are overcome by meta-heuristics methodologies. Many methodologies are implemented on economic dispatch like GA, ABCO and PSO [2–6]. But proposed DPSO when varying all variants of PSO has not implemented yet.

2 Nomenclatures

a_{ij}, b_{ij}, c_{ij}	The cost coefficients of the jth generator in area i (Rs/h), (Rs/h MW^{-1}), (Rs/h MW^{-2})
C_1, C_2	Acceleration coefficients for best and social experience of PSO
C_{1b}, C_{1p}	Acceleration coefficients for best and preceding experience
e_{ij} and f_{ij}	The valve-point effect coefficients of the jth generator in area i (Rs/h, MW^{-1})
gbest^t	Best particle during tth iteration
grms^t	Root mean square experience of the swarm during tth iteration

itr	Current iteration count
$\text{itr}_{\max}$	Maximum iteration count
$\text{itr}_{\min}$	Minimum iteration count
k	Ratio of dynamic cognitive and social acceleration coefficients
k_w	Ratio of maximum and minimum bound of the inertia weight
M	Number of areas
N_{Gi}	Number of generating units in the system in area i
N_{Gi}	Number of generating units in the ith area
$P\text{best}_n$	Best position of nth particle achieved based on its own experience
PD	Total real power demand of the system (MW)
$P_{\text{D}i}$	Total real power demand of area i (MW)
$P_{\text{G}i}$	Total real power generation in area i (MW)
$P_{\text{G}ij}$	Real power output of the jth generator in area i (MW)
$P_{\text{G}ij}^{\min}/P_{\text{G}ij}^{\max}$	Minimum/maximum generation limits of jth generator in area i (MW)
$P_{Tim}^{\min}/P_{Tim}^{\max}$	Minimum/maximum tie-line power limit from area i to area m (MW)
$p\text{preceding}_n$	Preceding position of nth particle achieved based on its just previous experience
P_{Tim}	Tie-line real power flow from area i to area m (MW) rand1 () and rand2 () random numbers in [0,1]
V_n^t	Velocity of nth particle at tth iteration
W	Inertia weight
$W_{\min}/W_{\max}$	Minimum/maximum value of inertia weight
Δt	Time step (s)
ζ_1 and ζ_2	Exponential constriction functions
η	Ratio of current and maximum iteration count
η_t	The value of g at which cognitive and social behaviour equalizes
μ	Constant
μ_1, μ_2	Coefficients of exponent terms

2.1 Problem Formulation

Cost function of generator is generally quadratic when valve-point loading is not considered. Large turbine generator usually has no. of fuel admissions valve that are operated such that to meet the increased generation. Objective function of MAED is stated as:

$$\text{Minimize } F(P_{\text{G}ij}) = \sum_{i=1}^{M}\sum_{j=1}^{N_{Gi}}\left(a_{ij} + b_{ij}P_{\text{G}ij} + c_{ij}P_{\text{G}ij}^2\right) + \left|e_{ij}\sin\left(f_{ij}\left(P_{\text{G}ij}^{\min} - P_{\text{G}ij}\right)\right)\right| \quad (1)$$

Subject to following constraints,

Area power balance constraints:

Total power generation of all generators is equal to the demand $P_{\text{D}i}$.

$$\sum_{j=1}^{N_{Gi}} P_{\mathrm{G}ij} = P_{\mathrm{D}i} + \sum_{m,i\neq m} P_{Tim} \quad i \in \{1, 2, \ldots M\} \tag{2}$$

Generator constraints:

$$P_{\mathrm{G}ij}^{\min} \leq P_{\mathrm{G}ij} \leq P_{\mathrm{G}ij}^{\max} \tag{3}$$

Tie-line constraints:

$$-P_{Tim}^{\min} \leq P_{Tim} \leq P_{Tim}^{\max} \tag{4}$$

3 Solutions Technique

3.1 Proposed DPSO

PSO mainly depends on three components, i.e. inertial weight, cognitive and social influence factors.

Each particle velocity and position is updated by the equations that are

$$\begin{aligned} \upsilon_n^{t+1} &= W \times \upsilon_n^t + e_1 \times c_{1b} \times r_1() \times (p\mathrm{best}_n - x_n^t) + (1 - e_1) \\ &\times c_{1p} \times r_2() \times (x_n^t - p\mathrm{preceding}_n) + e_2 \times c_2 \times r_3() \\ &\times (g\mathrm{best}^t - x_n^t) + e_2 \times c_2 \times r_4() \times (g\mathrm{rms}^t - x_n^t) \\ S_n^{t+1} &= S_n^t + V_n^{t+1} \times \Delta t \end{aligned} \tag{5}$$

where Δt is time generally taken as 1 s.

Inertia weight equation is:

$$W = W_{\min} + \frac{(W_{\max} - W_{\min}) \times (\mathrm{itr}_{\max} - \mathrm{itr})}{\mathrm{itr}_{\max}} \tag{6}$$

In proposed DPSO, the inertial weight W is changed by proposing an exponentially decomposed function to check trade-off between overall explorations and ceil exploitation of swarm. The preceding expertise of particle $p\mathrm{preceding}_n$ is taken to enrich the cognitive part by studying just older experience, whereas total experience of group of swarms is embedded in $g\mathrm{rms}^t$ to enrich the social part of particles. Further, dynamic acceleration coefficients have been introduced using constriction functions e_1 and e_2 dynamically to regulate the cognitive and social nature of birds. These modifications are explained below:

- Updating inertia weight: The trend of linear module of inertial weight is succeeded to solve ED problems using PSO by many researchers. In the suggested method, inertial weight is permitted to vary an exponential decomposed function and module suggested to modernize the inertial weight which is controlled by the following relation:

$$W = \mathrm{e}^{(-\eta \ln k_w)}$$

$$k_w = \frac{w_{\min}}{w_{\max}}$$

where $\eta = \mathrm{itr}/\mathrm{itr}_{\max}$ and k_w is selected within minimum and maximum bounds of inertial weight. In this study, value of k_w is proportion of max and min condition of inertial weight.

Updating preceding experience: The cognitive behaviour was split in by considering also the bad experience in addition to good idea of particle and provides certain more diverseness, but it results in poor local exploitation unless supported by a local random search. Therefore in existing method, the plan of preceding experience is evoked where existing fitness of every particle is compared with its fitness value in that preceding iteration, and if it is found less, it will be treated as the preceding experience. The past experience of the random particle produces less variety in comparison with the worst experience resulting in better exploration and exploitation of entire quest space without the aid of any additional local random search or else.

- Updating RMS experience: In PSO, only local and global best positions are transparent to other particles. This poor communication among particles may lead to lack of diversity and thus results in poor performance, especially while dealing with large dimensional problems. One way to improve communication of particles is to add RMS component of all particles' velocities in the measured equation. These results in overall sharing of data and particles gained from the discoveries and older experience of each escort during their search.
- Dynamic regulation of acceleration coefficients: Governance via aid of static acceleration coefficients of social and cognitive nature of particles is done in conventional PSO. It is suggested by many researchers that in order to regulate particle velocity dynamic control must be experienced. In the recent work, with the introduction of two exponential constrictions function e_1 and e_2 acceleration coefficients are made dynamic. The cognitive and social conduct of swarm is regulated by the aid of constriction functions while this regulation is taking place. Velocities of particles are limited during their flight and are as follows:

$$e_1 = \mathrm{e}^{(-\mu_1 \eta)}$$

$$e_2 = k \cdot \mathrm{e}^{(\mu_2 \eta)}$$

$$k = \frac{e_1 \cdot c_{1b}}{e_2 \cdot c_2}$$

where k is proportion of conducted dynamic cognitive and social acceleration coefficients. For same values of these coefficients at $\eta = \eta_t$.

These variations control the conventional PSO that regulates particles velocity, within predefined limits without any extra formulation is bang in various enhanced performance of PSO, yet preserving variety because of stochastic nature of cognitive and social behaviours of swarm.
The coefficients of the exponents μ_1 and μ_2 are selected as 5 and 3.9, as beyond that the term $e^{-\mu_1 \eta}$ is not perceptible in ending of search. Further, the most applicable value of η_t is obtained as 2/3 after experimentations.

Thereafter, performing 1000 independent trials of DPSO we get average fuel cost, and on the basis of this average cost, we obtain optimal solution.

3.2 *Methodology of DPSO*

The solution of an MAED problem is the set of most optimal generations for the desired objective bounded by certain operational constraints. In the proposed PSO, the particles are encoded in real numbers as the set of current generations in MW, as shown in Figs. 1 and 2.

(a) Particle encoding and initialization: The initial population is randomly created with predefined number of particles to maintain diversity. Each of these particles satisfies the problem constraints already defined. Infeasible particle, whenever appeared, is corrected by employing a constrained handling algorithm as described later in the section. The fitness of each particle is evaluated using Eq. (1), and then *p*best, *p*preceding, *g*best and *g*rms are initialized. The initial velocity of particles is assumed to be zero.
(b) Constrained handling: The velocity and position update may create infeasible solutions. In profound research, constrained handling algorithm is used to correct the infeasible individuals by the help of feasible ones. Using this process, power is adjusted within its limits and tie-line variables as given in Eqs. (2) and (3), respectively. If the generated power is within its min or max generation level, then set it as its lower or higher bound limits as in Eq. (4). Similarly, if the transfer of real tie-line powers from area *i* to area *k* exceeds its limit then the tie-line power is fixed at tie-line limits as mentioned in Eq. (4) for security consideration. The power balance error is calculated by Eq. (2). The error is equally divided among all generators, and the procedure is iterated till it disrates to a predefined mismatch value *e*. In this work, *e* is taken as 0.001.
(c) Solution preservation and termination criterion: In stochastic-based algorithms like PSO, the solution with the good fitness in the present iteration may be lost in the next iteration. Particle with best fitness is *k* hold preserved for upcoming iteration. The algorithm stops when we reached to maximum iteration or all particles converges to a single position.

PG11	PG12	…	PG1j	…	PG1N
PG21	PG22	…	PG2j	…	PG2N
…	…	…	….	…	…
PGi1	PGi2	…	PGij	…	PGiN

Particle encoding

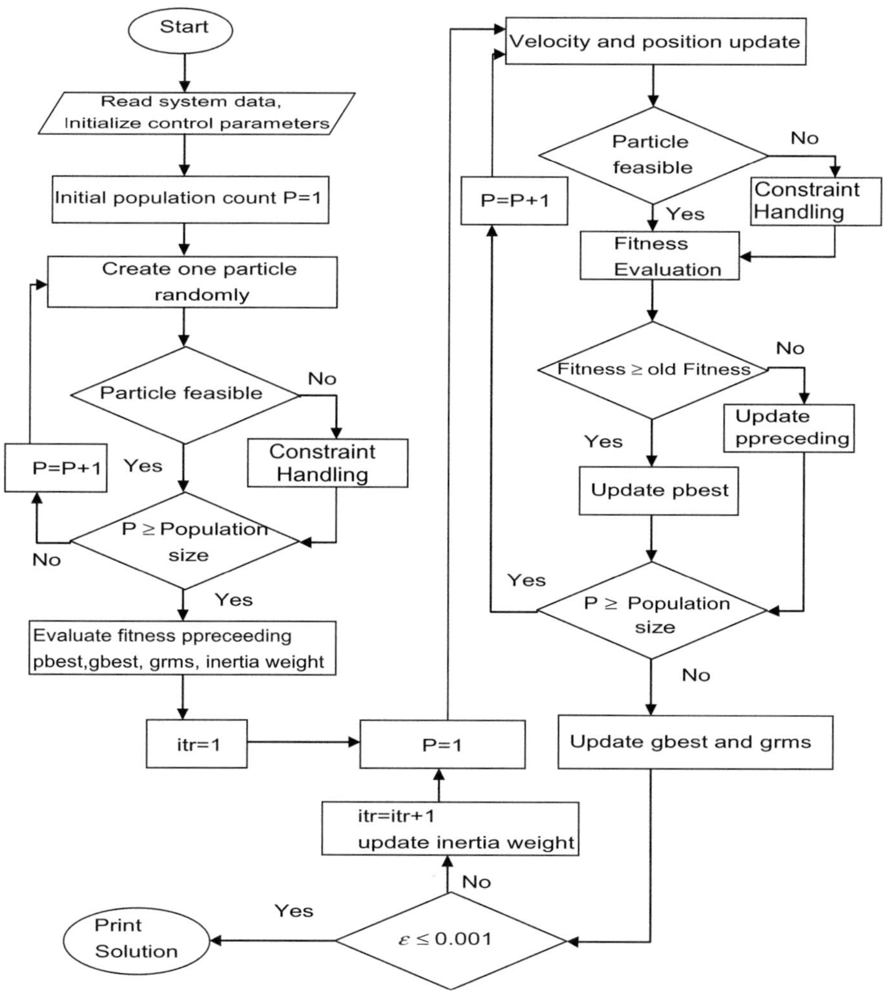

Fig. 1 Flowchart of MAED using DPSO

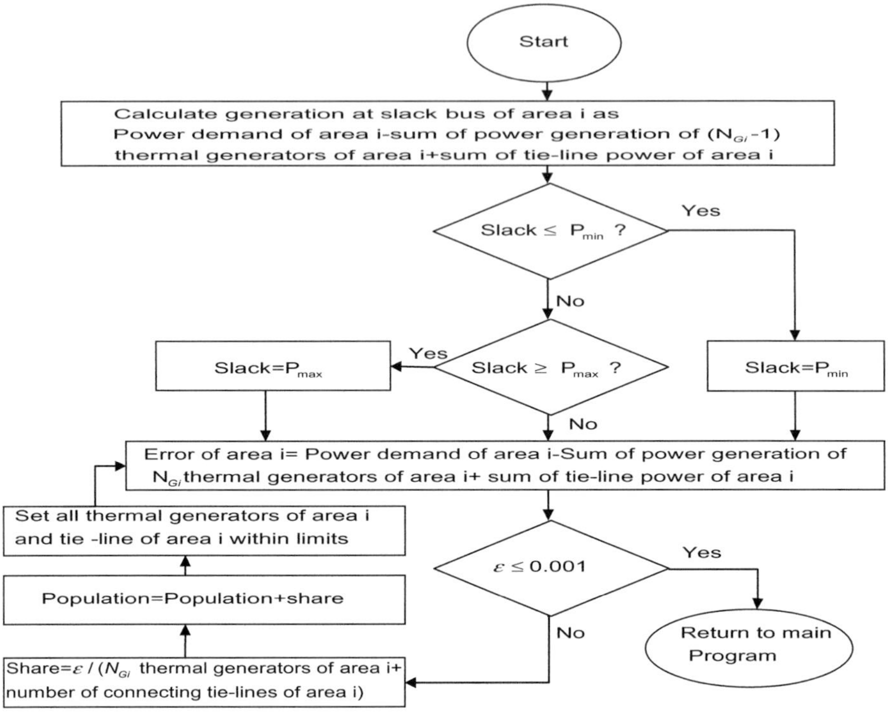

Fig. 2 Constraint handling algorithm of DPSO

3.3 Particle Encoding and Initialization Methodology Algorithm

Step 1: Initialize the control parameters and enter the system data.
Step 2: Initialize new particle randomly.
Step 3: If the current particle is not feasible, run constraint handling algorithm for it.
Step 4: Else go to Step 5.
Step 5: Increase population count by 1. If population count is less than its maximum, go to Step 2.
Step 6: Else go to Step 7.
Step 7: Evaluate fitness by (Eq. 1), pbest, ppreceding,*g*rms, inertia weight, constriction function by Eqs. (5) and (6).
Step 8: Initialize iteration count.
Step 9: Repeat Step 2 and Step 3. Update ppreceding for the current particle. Then repeat Step 5.

Step 10: Update gbest andgrms.
Step 11: Increase iteration count by 1. If iteration count is less than its maximum repeat Step 9.
Step 12: Print final results.

4 Test Systems and Results

4.1 Test System 1: Single Area Problem

Test system 1 simply has one area having no tie line. Table 1 shows coefficients of three generators with maximum and minimum limits of power, whereas Table 2 shows cost when demand is varying of test system 1, and its results are compared with classical methods, i.e. lambda iteration method (Fig. 3).

Test system 2 simply has one area having a tie line. Table 3 shows coefficients of four generators with maximum and minimum limits of power, whereas Table 4 shows cost when demand is varying of test system 2, and its results are compared with base paper (Fig. 4; Tables 5, 6, 7 and 8).

Table 1 Cost coefficients of three generators [1]

a	b	c	P_{min}	P_{max}
561	7.92	0.001562	150	600
310	7.85	0.00194	100	400
78	7.97	0.00482	50	200

Table 2 Cost of three generator units according to demand

P_D (power demand) (MW)	PSO (Rs/h)	Classical method (Rs/h)
850	8194.4	8194.4
1000	9583.2	9583.1

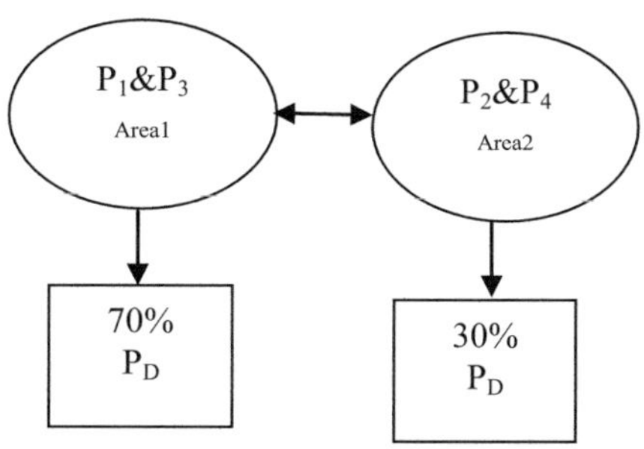

Fig. 3 Test system 2: two areas, four generator units and a tie line

Table 3 Cost coefficients of four generators

a	b	c	P_{min}	P_{max}
561	7.92	0.001562	150	600
310	7.85	0.00194	100	400
78	7.97	0.00482	50	200
250	7.5	0.00181	70	340

Table 4 Cost of four generator units and its tie-line value

P_D (power demand) (MW)	Tie-line limit (MW)	PSO-TVAC (Rs/h)
1000	200	9520.4
1120	200	10,605

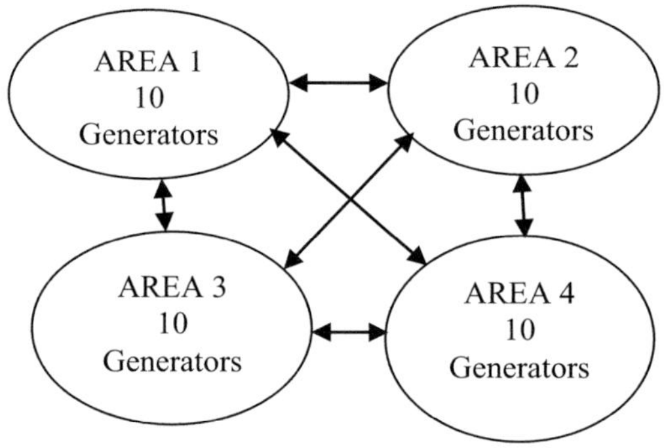

Fig. 4 Test system 3: 4 areas, 40 generation units, 6 tie-lines limit. Each area consists of ten generators with valve-point loading and is connected with three tie lines [7]

Table 5 Results of four generator units and two areas

Power (MW)	$P_D = 1000$ (MW) (PSO-TVAC)	$P_D = 1120$ (MW) (PSO-TVAC)
P_1	381.73	444.95
P_2	195.58	215.93
P_3	118.3	139.05
P_4	304.39	320.07
Tie-line power flow (MW)	199.97	200
CPU time In seconds	0.37159	0.27059

Test system 3 simply has four areas having six tie lines. Table 9 shows power of each generator, whereas Table 11 shows tie-line limit of each area of test system 3, and its results are compared with base paper (Table 10).

Table 6 PSO-TVAC parameters [7]

Parameters	C_1	C_2
Initial value	1.8	0.2
Final value	0.2	1.9

Table 7 Statistical analysis for 25 runs of the system

Method	Average cost (Rs/h)	Best cost (Rs/h)	Worst cost (Rs/h)	Mean time (in sec)	Power violation	Std. dev
PSO-TVAC	10605.144	10,605	10605.34	0.264	0.00	0.077
ABCO [8]	10617.5	10608.6	10664.3	4.35	0.00	27.83

Table 8 Parameters taken into account to deal with test system 3 [8]

Parameters	Value
Total power demand (MW)	10,500
Tie-line limit (MW)	200/100
Area load demand (%)	15/40/30/15
w_{max}	0.9
w_{min}	0.1
c_{1b}	2
c_{1p}	0.5
μ_1	5
μ_2	3.9
η_t	2/3
K	4
itr_{min}/itr_{max}	1/100

4.2 Convergence Curve of Test System 3

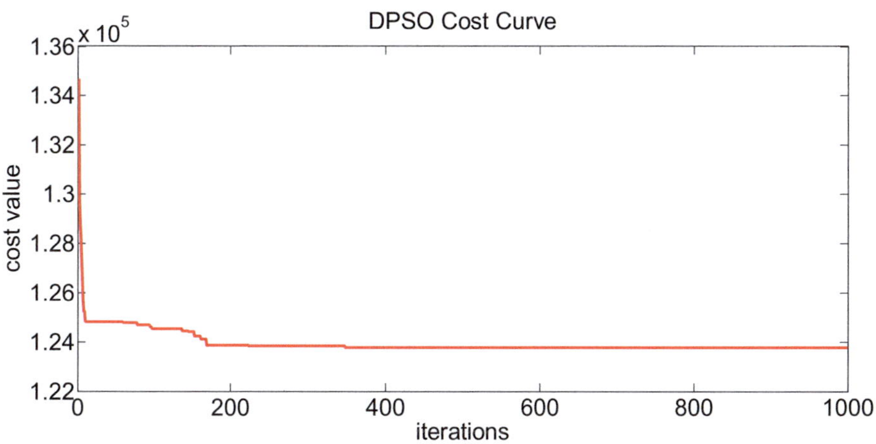

Table 9 Results of power of 40 generators by DPSO

Gen	DCPSO (MW)	Gen	DCPSO (MW)
P_1	110.5595	P_{21}	523.1689
P_2	110.5595	P_{22}	523.1689
P_3	116.5595	P_{23}	523.1689
P_4	186.5595	P_{24}	523.1689
P_5	93.5595	P_{25}	523.1689
P_6	136.5595	P_{26}	523.1689
P_7	260.2644	P_{27}	10
P_8	296.5595	P_{28}	10
P_9	296.5595	P_{29}	10
P_{10}	93.5595	P_{30}	47
P_{11}	94	P_{31}	190
P_{12}	94	P_{32}	190
P_{13}	125	P_{33}	190
P_{14}	486.610	P_{34}	168.744
P_{15}	486.610	P_{35}	168.744
P_{16}	486.610	P_{36}	168.744
P_{17}	486.610	P_{37}	110
P_{18}	486.610	P_{38}	110
P_{19}	536.610	P_{39}	110
P_{20}	536.610	P_{40}	320.744

Table 10 Statistical analysis for 25 runs of the system

Method	Average cost (Rs/h)	Best cost (Rs/h)	Worst cost (Rs/h)	Mean time (in sec)	Power violation
DCPSO	123790.9	123599.2	123,935	114.21	0.000
ABCO [8]	–	124009.4	–	126.934	0.000
DE	–	124544.1	–	134.8	0.000
EP	–	124574.5	–	144.5	0.000
RCGA	–	129911.8	–	160.5	0.000

Table 11 Tie-line results of test system 3 in both DCPSO

Tie-line limit (MW)	By DCPSO (MW)
1–2	107.5098
1–3	47.4277
1–4	7.8029
2–3	−186.610
2–4	−86.61
3–4	−73.1689

5 Conclusion

The conditions of today's market require the complex network to be considered as a set of separated, but interconnected areas and case are also very complex as it is characterized with huge number of networks, constraints, tie lines and loads. The research is successful in establishing a reliable, an efficient and fast heuristic search technique, DPSO and OPF to unravel MAED problems. The results obtained are substantially better both in the terms of cost and CPU time. MAED problem if high dimensional (as in the present case) also suffers with the case of local trappings when solved using any population-based algorithm. But as DPSO has improved the results significantly from the previous best results of powers, hence the problem is of high dimensionality and local trappings are assumed to be mitigated. The presented work can be extended to a larger problem of MAED which can further establish the validity of the present work and the DPSO technique. In future, the DPSO technique can be further modified within itself or it can be merged with another optimization technique, classical or meta-heuristic or otherwise, so that it further reduces the gap of obtained results and global best results which are not established as of now.

References

1. Wood, Wollenberg: Power Generation Operation & Control
2. Lee, K.Y., El-Sharkawi, M.A.: Modern Heauristic Search Techniques
3. Selvakumar, A.I., Thanushkodi, K.: A new particle swarm optimization solution to non-convex economic dispatch problems. IEEE Trans. Power Syst. **22**(1), 42–51 (2007)
4. Manoharan, P.S., Kannan, P.S., Baskar, S., Iruthayarajan, M.: Evolutionary algorithm solution and KKT based optimality verification to multi-area economic dispatch. Int. J. Electr. Power Energy Syst. **31**(7–8), 365–373 (2009)
5. Jadoun, V.K., Gupta, N., Niazi, K.R., Swarnkar, A.: Multi-area economic dispatch with reserve sharing using dynamically controlled particle swarm optimization. Electr. Power Energy Syst. **73**, 743–756 (2015)
6. www.al-roomi.org/multimedia/Economic_load_dispatch/4Units/40Units_ELD_TestSystem
7. Sharma, Manisha, Pandit, Manjaree, Srivastava, Laxmi: Reserve constrained multi-area economic dispatch employing differential evolution with time-varying mutation. Int. J. Electr. Power Energy Syst. **33**(3), 753–766 (2011)
8. Basu, M.: Artificial bee colony optimization for multi-area economic dispatch. Int. J. Electr. Power Energy Syst. **49**, 181–187 (2013)

An Analytical Hierarchy Process Based Approach for Effective Maintenance Prioritization of Power Transformers

Kavish Jain, Awin Gupta, Yog Raj Sood and Manisha Sharma

Abstract Transformer diagnosis and condition monitoring, literally a big deal for the utility professionals and is still posing a challenge to the researchers and domain experts of this area. It is clear that the accuracy of the diagnosis or maintenance process will more or less depend on the accuracy of the competence professional. Considering this as a constraint, lot of work has been done by eminent researchers using various artificial intelligence techniques like artificial neural networks, fuzzy logic and various algorithms for uplifting the accuracy of the diagnosis. Thus, a step ahead in this work, maintenance prioritization of power transformer has been carried out by using a Multiple Attribute Decision Making (MADM) technique, i.e., Analytical Hierarchy Process (AHP). Seven power transformers for which maintenance scheduling is to be done were considered as alternatives, and the results of various diagnostic tests such as furan analysis (2-FAL), total dissolved combustible gases (TDCG), Breakdown voltage (BDV), Dissipation Factor (DF), Acidity and Moisture content were considered as attributes for the corresponding transformer or alternative. Step-by-step calculations were done, and an optimal scheduling for these seven transformers was recommended considering the vulnerability state of insulation oil and criticality of the transformer.

Keywords Analytical hierarchy process · Prioritization · Power transformers
Condition monitoring

K. Jain (✉) · A. Gupta · Y. R. Sood · M. Sharma
Electrical Engineering Department, National Institute of Technology Hamirpur, Hamirpur, India
e-mail: kavishjain.23291@gmail.com

A. Gupta
e-mail: gupta.awin2105@gmail.com

Y. R. Sood
e-mail: yrsood@gmail.com

S. N. Singh et al. (eds.), *Advances in Energy and Power Systems*, Lecture Notes in Electrical Engineering 508, https://doi.org/10.1007/978-981-13-0662-4_15

1 Introduction

Power transformer is one of the most essential equipment in the modern power system, as it comprises more than 60% of the total investment in substations. In the rapidly expanding power system, the load on power transformer is increasing day by day to meet the end user needs, hence it has pushed most power utilities to assess the actual condition of transformers and adopt a proper maintenance strategy in case of failure [1–3]. At present, a majority of maintenance tasks in power system are carried out as a result of unexpected failure. Though the fault is cleared, the probability of occurring of these types of failures again is not reduced. Also, the money spent due to improper maintenance or over maintenance is about one-third of the total maintenance cost of the equipment. Hence, by implementing proper maintenance strategy, the overall running cost of the transformer can be reduced significantly.

At any site, maintenance of transformers is carried out by various condition monitoring experts having vast amount of knowledge. These maintenance experts rely more on their experience. However, even if they have sufficient amount of experience, at times it may not be possible for them to deal with the huge amount of condition monitoring data that they are provided with while carrying out the maintenance plan for a device like transformer [4, 5]. So, there is a need to develop a mathematical tool which provides us with the information to carry out the maintenance task in a structured and well-defined way.

In the past, several studies have been formulated related to power transformer condition assessment and life management techniques. These techniques include monitoring of oil test data such as Breakdown Voltage (BDV), Moisture Content (MC), Dissolved Gas Analysis (DGA), Dissipation Factor (DF), Acidity (ACI) and monitoring paper insulation through test like Degree of Polymerization which includes test like 2-furfuraldehyde (FAL) [6–12]. All these tests are conducted on timely basis or based on the condition of the transformer. However, till date no mathematical model has been developed that could take all the test data of oil and paper insulation mentioned above together and provide us the priority in which the maintenance schedule of a power transformer is to be carried out.

This paper introduces a technique called as Analytical Hierarchy Process (AHP), developed by Thomas L. Saaty in 1980 [13], to prioritize the transformers for maintenance by combining results of all the tests performed on oil and paper insulation mentioned above. A total of seven transformers were taken which are the alternatives and test results Moisture Content (MC), Acidity (ACI), Breakdown Voltage (BDV), Dissipation Factor (DF), Total Dissolved Combustible Gases (TDCG), and 2-Furfuraldehyde (2-FAL) were considered as attributes for maintenance prioritization. Step-by-step calculations were done and the final results were obtained.

Table 1 shows the test results of seven different transformers that were obtained from TIFAC-Core Lab, NIT Hamirpur.

Table 1 Test data of seven transformers

Alternatives	Attributes					
	MC (ppm)	ACI (mgKOH/g)	BDV (kV)	DF (%)	TDCG (ppm)	2-FAL (ppm)
Transformer-1(T-1)	22	0.203	70.7	0.169	1258	0.357
Transformer-2(T-2)	35	0.256	45.6	0.409	954	0.457
Transformer-3(T-3)	21.2	0.349	33.5	0.257	367	8.907
Transformer-4(T-4)	23.6	0.496	43.1	0.654	1483	7.76
Transformer-5(T-5)	36.7	0.344	42.4	0.657	1283	1.119
Transformer-6(T-6)	8.5	0.275	71.1	0.467	792	0.109
Transformer-7(T-7)	24	0.23	27	0.380	5638	7.40

2 Importance of Attributes

The attributes taken in the present work play a very significant role in defining the condition of a given transformer. The importance of each attribute is explained below.

2.1 Moisture Content (MC)

A huge proportion of transformer population is contaminated with moisture to some extent. The presence of moisture in transformer oil has significant effects on health of transformer such as accelerated rate of aging [14, 15], decrease in dielectric strength of oil, etc. The presence of moisture in oil also gives us the information about the condition of paper insulation. Hence moisture is one important factor in determining the condition of transformer.

2.2 Acidity (ACI)

Acidity, which is measured in mgKOH/g is another important parameter which provides us information about the deterioration of transformer insulation [15]. Extending the timely servicing periods of transformers increases the acidity, hence providing us knowledge about the condition of transformer.

2.3 Breakdown Voltage (BDV)

BDV is the measure of the ability of insulating oil to withstand voltage stresses [16]. The relevant standard for BDV test is ASTM D1816, and the test data under study has been obtained from this standard only [16]. Lower value of BDV accelerates the aging of transformers as it causes increased partial discharges and sparks.

2.4 Dissipation Factor (DF)

To estimate the electrical properties and the contamination of transformer oil, DF is evaluated. It also provides us about the information of the total power lost in the transformer oil during its operation. The total power dissipated is transferred in the form of heat energy to transformer oil, thereby causing the increase in temperature of oil, which in turn increases the temperature of transformer and help in aging.

2.5 Total Dissolved Combustible Gases (TDCG)

The gases evolved during the operation of transformer are known as Dissolved Combustible Gases. However, there is increase in the concentration of these gases when the fault is present inside the transformer or when the transformer is aged [17]. The gases obtained during the Dissolved Gas Analysis (DGA) are H_2, CH_4, C_2H_2, C_2H_4 and C_2H_6, and the summation of all these gases is called as Total Dissolved Combustible Gases.

2.6 2-Furfuraldehyde Content (2-FAL)

One of the most important parameters which is used to assess the condition of power transformer is 2-FAL, because it directly provides us the information about the condition of the solid insulation, a reduction in which is the main reason behind the shutdown of transformer and reaching of its end of life [3, 18].

3 Multiple Attribute Decision Making (MADM) Methods

Multiple Attribute Decision Making (MADM) method is a powerful tool which is used to solve problems where number of options or choices or alternatives exist [19]. Using MADM method, if we have information about the attributes of various alternatives, then we can arrive at a solution, i.e., best alternative or the rank of whole alternatives. This method is carried out with the help of decision table which contains information about alternatives (here transformers T-1, T-2, T-3, T-4, T-5, T-6, T-7), attributes (here MC, ACI, BDV, DF, TDCG, 2-FAL), relative importance or weights of attributes and performance measures of alternatives. In order to carry out MADM

Table 2 Value of attribute

Subjective measure of attribute	Assigned value
Exceptionally low	0.0
Extremely low	0.1
Very low	0.2
Low	0.3
Below average	0.4
Average	0.5
Above average	0.6
High	0.7
Very high	0.8
Extremely high	0.9

technique, all the elements in decision table must be normalized in the same units. The data shown in Table 2 is normalized using below relations.

For a beneficial attribute,

$$W_{(ij)\text{norm}} = \frac{W_{ij}}{W_{(j\text{max})}} \quad (1)$$

For a non-beneficial attribute,

$$W_{(ij)\text{norm}} = \frac{W_{(j\text{min})}}{W_{ij}} \quad (2)$$

where $W_{(j\text{max})}$ is the maximum value of attribute among all the alternatives for a beneficial parameter and $W_{(j\text{min})}$ is the minimum value of attribute among all the alternatives for a non-beneficial parameter. Attributes where maximum value is considered as better performance are called as beneficial attribute (e.g., BDV), and attributes where the minimum value is desired is called as non-beneficial attribute (e.g., MC, ACI, DF). But our aim is to prioritize the worst transformer first, so in our case the role of beneficial and non-beneficial attribute will be reversed. The final normalized data is shown in Table 3. This table is also considered to be a 7×6 Matrix, A_1 for future computations.

3.1 Analytical Hierarchy Process (AHP Method)

AHP method, developed by Saaty [13], is one of the most popular techniques used in decision making problems. In AHP method, determination of rank of alternatives is as follows,

Table 3 Normalized values of data, A_1

	MC	ACI	BDV	DF	TDCG	2-FAL
T-1	0.599	0.409	0.382	0.257	0.223	0.04
T-2	0.954	0.516	0.592	0.623	0.169	0.051
T-3	0.578	0.704	0.806	0.391	0.065	1
T-4	0.643	1	0.626	0.995	0.263	0.871
T-5	1	0.694	0.637	1	0.228	0.126
T-6	0.232	0.554	0.38	0.711	0.14	0.012
T-7	0.654	0.464	1	0.578	1	0.831

Table 4 Relative importance matrix, A_2

	MC	ACI	BDV	DF	TDCG	2-FAL
MC	1	6	3	5	1/3	1/4
ACI	1/6	1	1/5	1/3	1/7	1/8
BDV	1/3	5	1	4	1/5	1/6
DF	1/5	3	1/4	1	1/6	1/7
TDCG	3	7	5	6	1	1/2
2-FAL	4	8	6	7	2	1

- Determination of relative importance of each attribute with respect to a given objective:
  Here, using a scale of relative importance, first a pair-wise comparison matrix is constructed. In pair-wise comparison, if an attribute is compared with itself, then it is assigned the value 1, which means the diagonals elements in this matrix will always be 1. Later on, numbers 3, 5, 7, 9 corresponds to 'moderate,' 'strong,' 'very strong,' and extreme importance, respectively [13]. Suppose, we have N attributes, then we obtain an $N \times N$ square matrix, A_2. Here, a_{ij} denotes relative importance of attribute i with respect to attribute j. Also, $a_{ij} = 1$, if $i = j$ and $a_{ji} = 1/a_{ij}$. The pair-wise comparison, also known as relative importance matrix A_2, for all the parameters is listed in Table 4.
- After the formation of normalized matrix, the relative normalized weight W_j of each attribute is found out. It is done by calculating the geometric mean of the ith row and thereafter normalizing the geometric means of all the rows in the comparison matrix. This is done by use of following relations (Table 5):

$$P_j = \left[\prod_{j=1}^{N} a_{ij}\right]^{\frac{1}{N}} \tag{3}$$

$$W_j = P_j / \sum_{j=1}^{N} P_j \tag{4}$$

Table 5 Analytical hierarchy process method score, A_6

Attribute	P_j	$W_j = \frac{P_j}{\sum P} = A_3$	$A_4 = A_2 \times A_3$	$A_5 = A_4/A_3$
MC	1.399	0.152	0.977	6.427
ACI	0.241	0.026	0.174	6.692
BDV	0.778	0.085	0.564	6.635
DF	0.391	0.042	0.279	6.642
TDCG	2.608	0.285	1.815	6.368
2-FAL	3.728	0.407	2.611	6.415

Table 6 Random index (RI) values

Attributes	3	4	5	6	7	8	9	10
RI	0.52	0.89	1.11	1.25	1.35	1.4	1.45	1.49

- Obtain the matrices A_4 and A_5 by the relation $A_4 = A_2 \times A_3$ and $A_5 = A_4/A_3$.
- Maximum eigenvalue $\lambda_{\max}$, which is the average of matrix A_5, is determined by the formula:

$$\lambda\text{max} = \left[\sum_{j=1}^{N} A_{5_j}\right] / N \tag{5}$$

- Next, we obtain the value of Consistency Index CI using the relation below. CI gives us information about the deviation from consistency. A smaller value of CI is desired for accuracy of results.

$$\text{CI} = \frac{\lambda\text{max} - N}{N - 1} = 0.105 \tag{6}$$

- Next, the Random Index (RI) value is found out for the number of attributes used in decision making with the help of Table 6. Rao [19] showing Random Index (RI) values. We have 6 attributes, so $N = 6$, and RI value $= 1.25$.
- Next step is the calculation of Consistency Ratio (CR), using the equation below. The acceptable value of CR is 0.1 or less, and it indicates if the relative importance matrix selection is properly done or not.

$$\text{CR} = \frac{\text{CI}}{\text{RI}} = \frac{0.105}{1.25} = 0.084 < 0.1$$

- Once the Consistency Ratio (CR) is evaluated and verified, the next step is the multiplication of the normalized value of each alternative with its corresponding weight value W_j, i.e., the 7×6 matrix A_1 is multiplied with 6×1 matrix A_3 to get a 7×1 column matrix A_6 as shown in Table 7. The value of the column matrix

Table 7 Analytical hierarchy process method score, A_6

Attributes	Final weights
Transformer-1	0.225
Transformer-2	0.305
Transformer-3	0.618
Transformer-4	0.65
Transformer-5	0.384
Transformer-6	0.157
Transformer-7	0.845

A_6 gives us the performance scores or the rank in which we have to prioritize the alternatives.

- Finally, arranging these performance scores in the descending order will give us the rank of all the seven transformers according to which they will be chosen for effective maintenance prioritization.
- According to the final scores, the condition of transformer 7 (T-7) is worst and transformer 6 (T-6) is the best.

4 Conclusion

In this paper, transformer selection for effective maintenance prioritization has been carried out using a Multiple Attribute Decision Making (MADM) technique called as Analytical Hierarchy Process (AHP). Seven transformers were taken which were considered as alternatives, and six attributes namely Moisture Content, BDV, Acidity, Dissipation Factor, Total Dissolved Combustible Gases, and 2-Furfuraldehyde were considered. Step-by-step calculations of the algorithm were done, and finally, all the seven transformers were ranked according to their final scores or weights. The rank of each alternative is shown in Table 8.

Table 8 suggests that Transformer 6 (T-6) needs first priority in maintenance followed by transformer 5 (T-5), transformer 3 (T-3), and transformer 2 (T-2), and so on.

The results came out to be fairly accurate because the Consistency Index was less than 0.1%. Also, when the values of all the attributes given in Table 1 for all the transformers was checked according to the ASTM and CIGRE standards, it was seen that for transformer 7, all the values were exceeding and were beyond limits. Hence, we can say the results were fairly accurate.

Table 8 Rank of each alternative using AHP

Rank	T-1	T-2	T-3	T-4	T-5	T-6	T-7
AHP	7	4	3	5	2	1	6

Acknowledgements The authors are thankful to the authorities of TIFAC-CORE Centre of NIT Hamirpur (HP), India, for providing the necessary data for the transformers operating in Himachal Pradesh.

References

1. Muhamad, N.A.B., Bashir, N., Abuja, N., Alghamdi, A.S., Suleiman, A.A.: Asset management through effective transformer diagnostics & condition monitoring. In: IEEE International Conference on Power and Energy (PECon), Dec 2012, Kota Kinabalu Sabah, Malaysia, pp. 212–216
2. Wang, M., Srivastava, K.D.: Review of condition assessment of power transformers in service. IEEE Elect. Insul. Mag. **18**(6), 12–25 (2002)
3. Singh, J., Sood, Y.R., Jarial, R.K.: Condition monitoring of power transformers—bibliography survey. IEEE Elect. Insul. Mag. **24**(3), 11–25 (2008)
4. Ratnayake, R.M.C.: Modeling of asset integrity management process: a case study for computing operational integrity preference weights. Int. J. Comput. Sys. Eng. (IJCSysE) **1**(1), 3–12 (2012)
5. Antosz, K., Stadnicka, D.: Evaluation measures of machine operation effectiveness in large enterprises: study results. Eksploatacja i Niezawodność – Maint. Reliab. **17**(1), 107–117 (2015)
6. Zurich: ABB Service Handbook for Transformers, 2nd edn. ABB Management Service, Ltd., Switzerland (2007)
7. Wang, M., Srivastava, K.D.: Review of condition assessment of power transformers in service. IEEE Electr. Insul. Mag. **18**(6), 12–25 (2002)
8. Saha, T.K.: Review of modern diagnostic techniques for assessing insulation condition in aged transformers. IEEE Trans. Dielectr. Electr. Insul. **10**(5), 903–917 (2003)
9. Hjartarson, T., Otal, S.: Predicting future asset condition based on current health index and maintenance level. Presented at 11th IEEE Conference on Transmission & Distribution Construction, Operation and Live-Line Maintenance, Albuquerque, NM, Oct 2006
10. Naderian, A., Cress, S., Peircy, R.: An approach to determine the health index of power transformers. In: Proceedings of IEEE International Symposium on Electrical Insulation, June 2008, Vancouver, Canada, pp. 192–196
11. CIGRE Working Group 05: An international survey of failures in large power transformers in service. Electra **88**, 21–48 (1983)
12. Höhlein, I., Kachler, A.J., Tenbohlen, S., Leibfried, T.: Transformer life management German experience with condition assessment. Contribution for CIGRE SC12/A2, June 2003
13. Saaty, T.L.: The Analytic Hierarchy Process. McGraw-Hill, New York (1980)
14. Emsley, A.M., Xiao, X., Heywood, R.J., Ali, M.: Degradation of cellulosic insulation in power transformers. Part 3: effects of oxygen and water on ageing in oil. Proc. Inst. Elect. Eng. Sci. Meas. Technol. **147**(3), 115–119 (2000)
15. Lundgaard, L.E., Hansen, W., Linhjell, D., Painter, T.J.: Aging of oil-impregnated paper in power transformers. IEEE Trans. Power Del. **19**(1), 230–239 (2004)
16. IEEE Guide for Acceptance and Maintenance of Insulating Oil in Equipment, IEEE Standard C57. 106-2006 (Rev. IEEE Standard C57. 106-2002), 2006, pp. 1–36
17. IEEE Guide for the Interpretation of Gases Generated in Oil-Immersed Transformers, IEEE Standard C57. 104-2008 (Revision of IEEE Standard C57. 104-1991), 2009, pp. C1–27

18. Van Bolhuis, J.P., Gulski, E., Smit, J.J.: Monitoring and diagnostic of transformer solid insulation. IEEE Trans. Power Del. **17**(2), 528–536 (2002)
19. Rao, R.V.: Decision making in the manufacturing environment: using graph theory and fuzzy multiple attribute decision making methods. Springer Science & Business Media (2007)

Designing of Controllers Using Chebyshev–Pole Clustering Approximants

Vartika Rao

Abstract This paper proposes a technique for designing a controller based on Chebyshev–Pole Clustering Approximation piping to unit step input. The proposed controller reduces the intricacy of the higher order controller. Chebyshev Polynomial series and Pole Clustering technique are used to derive the coefficients of numerator and the denominator polynomial, respectively. A mathematical example is presented to exemplify the projected method. The blueprint of the controller prepared for the lower order model will be well suited to the higher order interconnected network.

Keywords Chebyshev Polynomial · Controller design · Model order reduction (MOR) · Pole Clustering technique

1 Introduction

The complexity of modeling interconnected structures is extensive so we are in a need of smaller models. Reduced order models or smaller models are the proposed structure that is able to trace the response of an interconnected network correctly, considering essential facts. The first and foremost requirement is that these prototypes should be less than the real models. At least one way, their behavior must be comparable to the behavior of the original model. Order reduction is a familiar subject within the simulation, control, and optimization of multifaceted systems. The mathematical models used in these computations often results in large-scale systems. We imagine that the original model is previously derived by physical laws and assumptions.

This technique focuses on retaining the essential features of an interconnected network. It can be assumed that the basic features are already present in the lower order model which has been derived from the higher version. Accuracy and precision are the main points which should be kept under consideration while performing this process.

V. Rao (✉)
IIT (ISM), Dhanbad, India
e-mail: raovartika79@gmail.com

S. N. Singh et al. (eds.), *Advances in Energy and Power Systems*, Lecture Notes in Electrical Engineering 508, https://doi.org/10.1007/978-981-13-0662-4_16

Number of MOR techniques for different type of systems such as linear continuous and discrete time systems have been introduced by C. B. Vishwakarma using Pade Approximation [1], Routh Approximation [2] for linear time-invariant system (LTI), time moment matching methods [3], error minimization method [4] and a method based on Chebyshev Polynomial techniques [5, 6]. Apart from these methods, some methods have been proposed which are composite forms of more than one method [7–9]. A composite method based on Pole Clustering and Chebyshev Polynomial technique of obtaining stable reduced order model has been proposed in 2012 [10].

Designing of controller for large-scale systems is a problem to be considered. As rise in the order of the system is noticed, then the intricacy of the designing of controller grows. To handle this issue, there is a need of good approximation of higher order model which can be obtained by different techniques mentioned above. Now the obtained lower order model can be used in the formation of a controller which when applied to the higher order original system provides satisfactory response.

Controller design and concept approaches have been presented by Anderson and Liu [11], Hyland [12, 13]. In [14], the method balanced realization has been discussed. The basic avenues can be divided into direct and indirect one as discussed by Hyland [12, 13]. A controller design based on ISE minimization technique has been discussed in [15].

The focus is on approximating the original higher order model by a lower order model using Chebyshev–Pole Clustering Approximant [10] and then the actual higher order plant is connected to the controller designed for this lower order prototype.

2 Background

Consider a system function of an nth-order linear SISO higher order system is:

$$G_{\text{org}} = \frac{A_0 + A_1 s + \cdots + A_{n-1} s^{n-1}}{B_0 + B_1 s + \cdots + B_n s^n} \tag{1}$$

The system function of the lower order model ($r < n$) is:

$$G_{\text{red}} = \frac{\widetilde{A}_0 + \widetilde{A}_1 s + \cdots + \widetilde{A}_{r-1} s^{r-1}}{\widetilde{B}_0 + \widetilde{B}_1 s + \cdots + \widetilde{B}_r s^r} \tag{2}$$

2.1 Pole Clustering

The calculation of cluster center is proposed by Vishwakarma [16] by using inverse distance measure principle. Let r real poles in one cluster group and P_c be the cluster center based on IDM criterion

$$P_c = \left\{ \left(\sum_{i=1}^{r} \left(\frac{1}{hi} \right) \right) \div r \right\}^{-1} \tag{3}$$

where P_c is cluster center of r real poles h_1, h_2, h_3, h_r of the original cluster. Assume r real pair of complex conjugate poles be $(h_{1R} + jh_{1I})$, $(h_{2R} + jh_{2I}) \ldots (h_{rR} + jh_{rI})$

$$h_r = \left\{ \left(\sum_{i=1}^{r} \left(\frac{1}{hir} \right) \right) \div r \right\}^{-1} \tag{4}$$

$$h_I = \left\{ \left(\sum_{i=1}^{r} \left(\frac{1}{hI} \right) \right) \div r \right\}^{-1} \tag{5}$$

While calculating the centers of cluster by inverse distance measure principle, the cluster center for real and unreal poles should have separate calculations and retention of the poles on imaginary axis should be done.

The denominator of the reduced model can be calculated by following steps:

1. Let all the calculated cluster centers are real, denominator comes as:

$$\widehat{D}r(s) = \prod_{i=1}^{r} [s - (P_c)_i]$$
$$= \text{Denominator of the Reduced order Model} \tag{6}$$

2. Assume the calculated cluster centers are complex conjugate, denominators come as:

$$\widehat{D}r(s) = \prod_{i=1}^{r/2} [s - (h_R + jh_I)][s - (h_R - jh_I)] \tag{7}$$

3. Let one cluster center is real and the remaining one is complex conjugate, denominator comes as:

$$\widehat{D}r(s) = (s - P_1) \prod_{i=2}^{(r+1)/2} [s - (h_R + jh_I)][s - (h_R - jh_I)] \tag{8}$$

2.2 *Chebyshev Polynomial*

We can define the first kind of this polynomial as $Tn(x)$, i.e., x of degree n, defined by the relation

$$Tn(x) = \cos n\theta \text{ when } x = \cos\theta$$

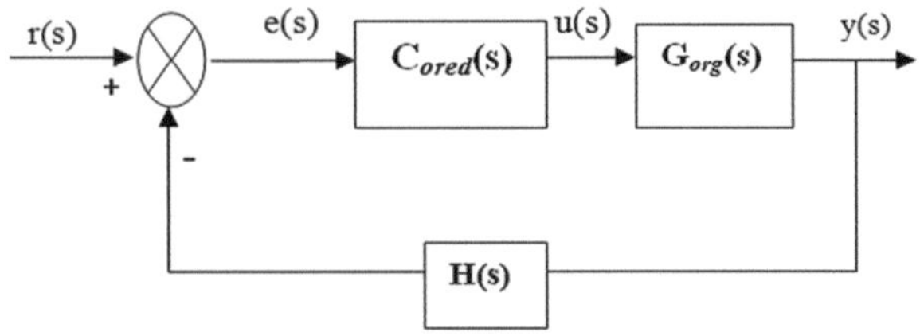

Fig. 1 Closed-loop control system

Condition: The range of the variable x is the interval $[-1, 1]$, and then the range of the corresponding variable θ can be taken as $[0, \pi]$. These ranges are traversed in opposite directions, since $x = -1$ corresponds to $\theta = \pi$ and $x = 1$ corresponds to $\theta = 0$.

It is well known that cos $n\theta$ is a polynomial of degree n in cos θ, and indeed we are familiar with the elementary formulae

$$\cos 0\theta = 1, \cos 1\theta = \cos\theta, \cos 2\theta = 2\cos 2\theta - 1,$$
$$\cos 3\theta = 4\cos 3\theta - 3\cos\theta, \cos 4\theta = 8\cos 4\theta - 8\cos 2\theta + 1$$

Now, that the first few Chebyshev Polynomials are:

$$T_0(x) = 1, T_1(x) = x, T_2(x) = 2x^2 - 1,$$
$$T_3(x) = 4x^3 - 3x, T_4(x) = 8x^4 - 8x^2 + 1$$

3 Controller Design

Let us think about the control system [12] as shown in Fig. 1. $G_{\text{org}}(s)$ and $H(s)$ are the given transfer function of the controller. $C_{o(s)}$ has to be developed which gives the desired response of the closed-loop system. An indirect method is applied for the designing of controller $C_{o(s).}$ The model parameters of a reference (desired) model is assumed, and it is used for the designing and obtaining the closed-loop transfer function of the controller.

The overall closed-loop system function:

$$G_{foo} = \frac{C_{o(s)}G_{\text{org}(s)}}{1 + C_{o(s)}G_{\text{org}(s)}H_{(s)}} \tag{9}$$

So,

$$G_{\text{des}}(s) = \frac{C_{o(s)}G_{\text{org}(s)}}{1 + C_{o(s)}G_{\text{org}(s)}H_{(s)}} \tag{10}$$

Using Eq. (2), the expression for controller comes out to be:

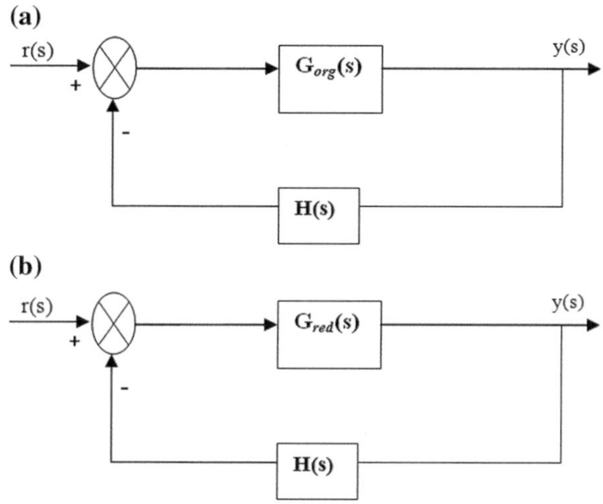

Fig. 2 **a** Closed-loop system, **b** reduced order model of the system

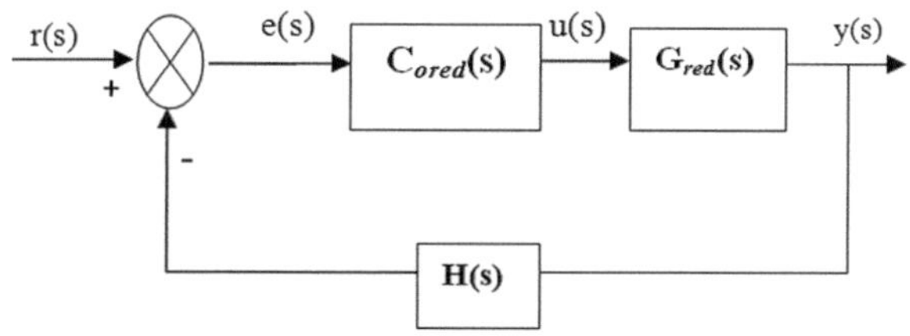

Fig. 3 Closed-loop control with $C_{ored}(s)$ and $G_{\mathrm{red}}(s)$

$$C_0(s) = \frac{G_{\mathrm{des}}(s)}{G_{\mathrm{org}}(s)[1 - G_{\mathrm{des}}(s)]} \tag{11}$$

The equation for the reduced order controller comes out to be:

$$C_{\mathrm{ored}}(S) = \frac{G_{\mathrm{des}}(s)}{G_{\mathrm{red}}(s)[1 - G_{\mathrm{des}}(s)]} \tag{12}$$

Chebyshev–Pole Clustering Approximation method is used to obtain reduced order model. $G_f(s)$ is known original higher order transfer function, and $G_{\mathrm{red}}(s)$ is the approximated lower order model. Both the models are shown in Fig. 2, where $H(s)$ is the feedback common in both the figures.

The closed-loop control structure of lower order model with reduced order is given in Fig. 3.

The complete closed-loop transfer function in Fig. 2 is given below:

$$G_{frr}(s) = \frac{C_{\mathrm{ored}}(s)G_{\mathrm{red}}(s)}{1 + C_{\mathrm{ored}}(s)G_{\mathrm{red}}(s)H(s)} \tag{13}$$

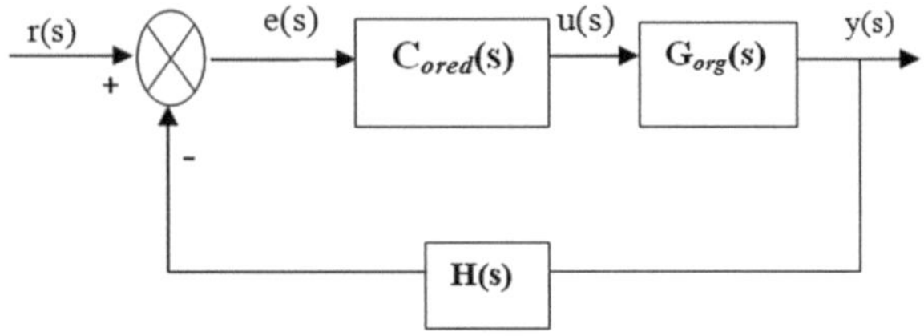

Fig. 4 Closed-loop control $C_{ored}(s)$ and $G_{org}(s)$

The closed-loop control structure of the given higher order plant connected with lower order controller is given in Fig. 4.

The complete closed-loop transfer function in Fig. 4 is given below:

$$G_{fro}(s) = \frac{C_{ored}(s)G_{org}(s)}{1 + C_{ored}(s)G_{org}(s)H(s)} \tag{14}$$

The above method shows that how a reduced order controller, i.e., $C_{ored}(s)$, is derived, and an example is shown in the next section.

4 Mathematical Example

Assume a fourth-order system described by the transfer function as:

$$G_{org}(s) = \frac{s^3 + 7s^2 + 24s + 24}{s^4 + 10s^3 + 35s^2 + 50s + 24} \tag{15}$$

Select a model which satisfies the control requirements. In this example, a standard second-order transfer function is chosen with damping ratio $\epsilon = 0.7$ and natural frequency $w_n = 1.5$ rad/s.

$$G_{des}(s) = \frac{2.25}{s^2 + 2.1s + 2.25} \tag{16}$$

Thus, transfer function $G_{ored}(s)$ becomes by using Chebyshev–Pole Clustering Approximation [10]:

$$G_{red}(s) = \frac{6.397 + 1.06s}{s^2 + 7.99s + 8.8885} \tag{17}$$

$$G_{frr}(s) = \frac{2.25s^3 + 58.77s^2 + 394.9506s + 316.21095}{s^5 + 28.22s^4 + 232.635s^3 + 567.92876s^2 + 690.08082s + 316.21095} \tag{18}$$

$$G_{fro}(s) = \frac{2.5s^5 + 63.135s^4 + 586.552s^3 + 2490.485s^2 + 4299.075s + 2263.14}{2s^7 + 30.1s^6 + 339.256s^5 + 1802.923s^4 + 4824.57s^3 + 7509.16s^2 + 6411.67s + 2263.14} \tag{19}$$

The step responses of (15), (16), (18), and (19) are shown in Fig. 5a–c, respectively.

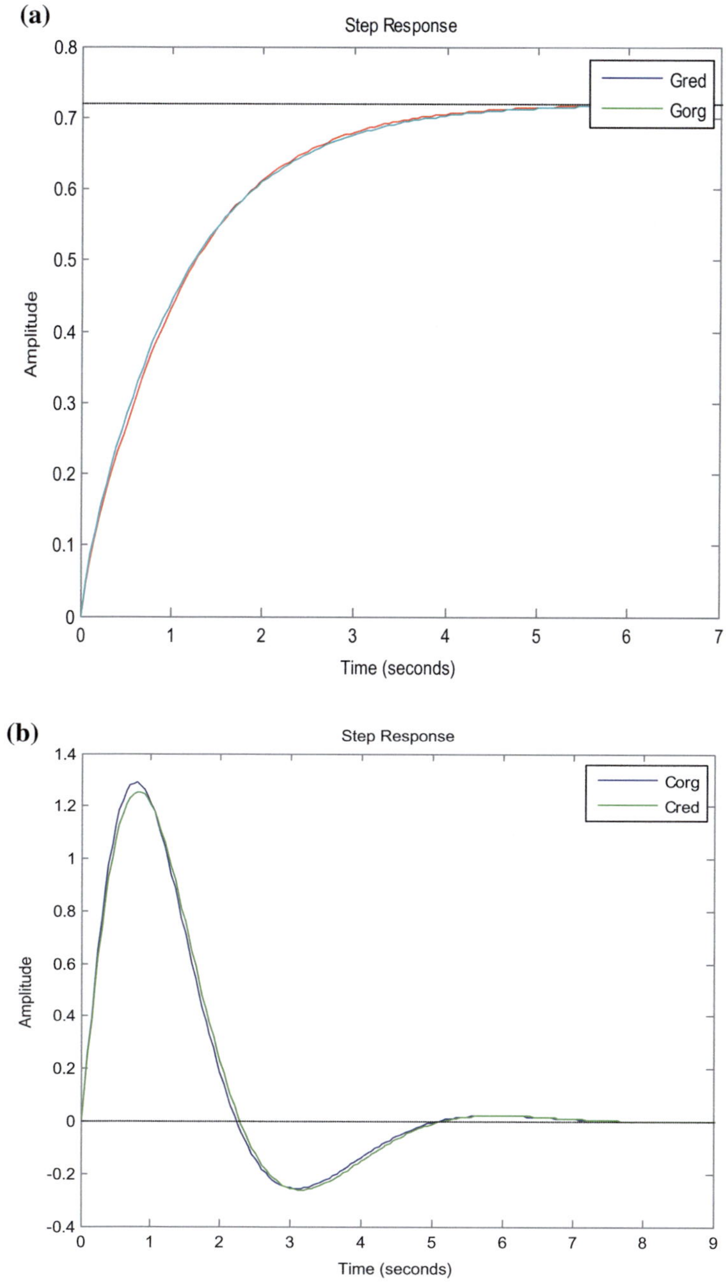

Fig. 5 **a** Step response, **b** step response, **c** step response

(c)

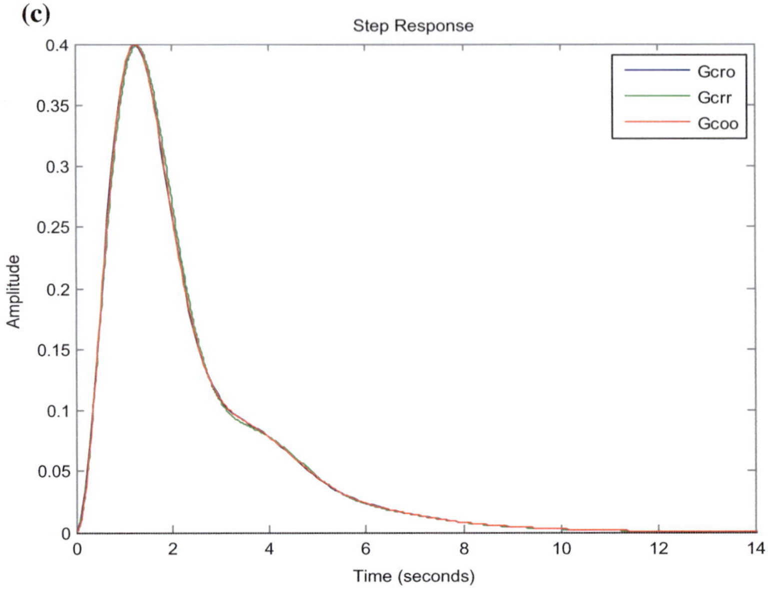

Fig. 5 (continued)

5 Conclusion

A method for controller design based on Pole Clustering and Chebyshev Polynomial technique method has been proposed. Higher order system has been approximated into the reduced order. A reference model $G(s)$ with assuming damping ratio $\epsilon = 0.7$ and natural frequency $w_n = 1.5$ rad/s has been chosen. Further, we can choose the reference model as per our requirement which gives flexibility to our design problem. Then this reduced order controller is implemented to the original higher order system, and the result illustrates that step response of the original plant with reduced order controller is a good approximant to the step response of original plant with higher order controller. Proposed controller is well suited for the reduced order system as well. The controller design technique can be compared to other methods so as to achieve a hybrid configuration to compensate the disadvantages of the present one.

References

1. Lavania, S., Nagaria, D.: Pade approximation based moment matching technique for model order reduction. Comput.-Aided Design Integr. Circ. Syst. IEEE Trans. **13**, 729–736 (1994). ISSN 0278-0070
2. Sachan, A., Rai, P., Kumar, P.: A conglomerating approach for model order reduction of continuous large scale systems. Int. J. Innov. Res. Sci. Eng. Technol. **5** (2016). ISSN 2319-8753

3. Huyuck, B., De Brabanter, J., Van Impe, J., De Moor, B.: Identification and modeling of a dynamical system. Submitted for Ecumict (2008)
4. Malekshahi, E., Mohammadi, S.-M.-A.: The model order reduction using LS, RLS and MV estimation methods. Int. J. Control Autom. Syst. (2014)
5. Baur, U., Feng, L., Benner, P.: Model Order Reduction for Linear and Nonlinear Systems: A System-Theoretic Perspective. Springer, Netherlands (2014)
6. Snyder, M.A.: Chebshev Methods in Numerical Approximation. Prentice Hall, Englewood Cliffs (1966)
7. Singh, V., Chandra, D., Kar, H.: Improved routh-pade approximants: a computer-aided approach. IEEE Trans. Autom. Int. J. Eng. Control **49**, 292–295 (2004)
8. Sobhy, M.I.: Applications of Pade approximants in electrical network problems. In: Graves-Morris, P. (ed.) Pade Approximants and Their Applications. Academic Press, NY (1973)
9. Chen, C.F., Shieh, L.S.: A novel approach to linear model simplification. Int. J. Control **8**, 561–570 (1968)
10. Vinay, P.S., Dinesh, C., Prateek, C., Member, IEEE.: Model Order Reduction of Continuous Time Systems using Pole Clustering and Chebyshev Polynomials. IEEE (2012). 978-1-4673-0455-9/12
11. Lamba, S.S., Gorez, R., Bandhyopadhyay, B.: New reduction technique by step error minimization for multivariable systems. (1988)
12. Dolgin, Y., Zeheb, E.: On Routh-Pad model reduction of interval systems. IEEE Trans. Autom. Control **48**(9) (2003)
13. Vishwakarma, C.B., Prasad, R.: Order reduction using modified pole clustering and Pade approximations. World Acad. Sci. Eng. Technol. **56**, 787–791 (2011)
14. Mukherjee, S., Mittal, R.C.: Model order reduction using response-matching technique. J. Franklin Inst. **342**(5) (2005)
15. Agrawal, S.K., Chandra, D., Irfan, A.K.: Controller design based on ISE minimization and dominant pole retention method. Int. J. Comput. Sci. Eng. (2010)
16. Vishwakarma, C.B., Prasad, R.: Clustering method for reducing order of linear system using Pade approximation. IETE J. Res. **54**(4) (2008)

Comparison: Matrix Converter, Cycloconverter, and DC Link Converter

Gurmeet Kaur and Sandeep Kumar

Abstract Converters are linear power electronics circuits that are used in wide area of electrical and electronics engineering field, but matrix converter has several advantages over traditional rectifier-inverter type power frequency converters. This paper reflects that matrix converter has several advantages over traditional rectifier-inverter type power frequency converters. The purpose of this work is to investigate the design and implementation of matrix converter for wind load. Also, a detailed explanation of the three matrix converter, cycloconverter, and DC link converter is presented. But the research is focused on direct AC–AC matrix converter topology. The principle characteristics of the AC drive can be fulfilled by matrix converter (MC) which is being increasingly used applications. It has the possibility to replace the conventional converter. The objectives of the research are to implement the MC to control speed and torque of PMS motor by using MATLAB Simulink and to design the converter for PMS motor application and to compare the performance of the motor by using matrix converter and common converters.

Keywords Matrix converter · Cycloconverter · DC link converter · Power electronics converters

1 Introduction

Generally, power electronics converters can be classified into various types like diode rectifiers, phase-controlled rectifiers, DC choppers, inverters, cycloconverters, matrix converters. In present days, a number of trends can be noticed in the area of converter design, namely traction control of electric vehicles, air conditioning, blowers and fans, pumps and compressors [1]. On the other hand, HVDC transmission, excitation system, conventional energy systems (solar, wind) still receive considerable attention

G. Kaur (✉) · S. Kumar
Electrical Department, Quantum School of Technology, Roorkee, India
e-mail: gurmeetkaur.ee@quantumeducation.in

S. Kumar
e-mail: sandeepkumar.ece@quantumeducation.in

S. N. Singh et al. (eds.), *Advances in Energy and Power Systems*, Lecture Notes in Electrical Engineering 508, https://doi.org/10.1007/978-981-13-0662-4_17

of many researches. A switching device which changes voltage frequency and phase while converting AC to AC is called an AC/AC converter or changer. In this category, the first converter group is direct frequency converter which converts AC to controlled AC. As mentioned previously, mainly they are of two types:

(1) DC link converters,
(2) AC–AC converters,
   (a) Cycloconverters,
   (b) Matrix converters.

2 DC Linked Converter

Due to their wide area of application, DC link converters are most commonly used converters. It can be used with a variety of loads, also it requires less maintenance, and it is highly efficient. This type of converter is stable in the condition of overloading and even in no loads due to the voltage stored in DC link [2]. Due to this reason, it can be detached from the load without damage in the above-mentioned conditions.

It also has an advantage that it can work in first, second, third, and fourth quadrant as per our need. Figure 1 shows the circuit of a DC link converter which consists of a rectifier, a DC link, and an inverter. The rectifiers indicated by V_{D1}, V_{D2}, ..., V_{D6} can work for three-phase as well as single-phase AC as per the requirement; this AC is converted to DC voltage whose order is proportional to the converted AC. The input of these converters was connected to the supply line via transformer in order to prevent the mains from harmonics and nonlinear distortions. An inductor is used at the output of the rectifier in order to work as a filter and remove or lower down the ripple content due to which the fault current is limited. The large capacitor C is used as an overvoltage protector. The circuit had a less amount of starting current due to the charging of the capacitor. The inverter switches are connected to the load side. Here, the flow of voltage is bidirectional due to which the converter can work in two out of the four quadrants. In order to make it work in all the four quadrants, regeneration of energy is done by the inverter back to the DC link; as a lacking factor, this regeneration could increase the power above the safe level so proper measures should be taken against this high voltage built up. To do so, we use a resistor R; this is connected across the capacitor C to absorb the extra regenerated energy.

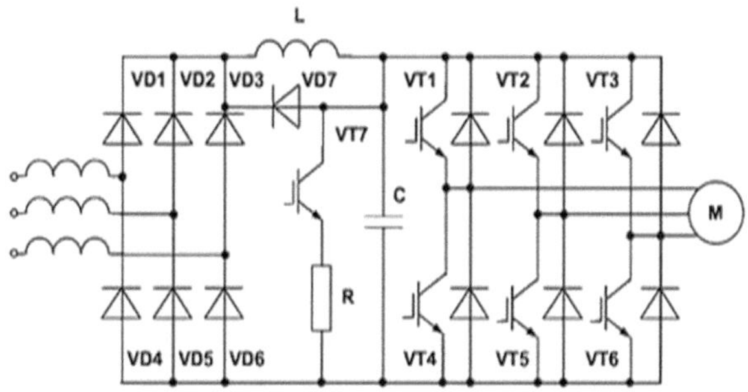

Fig. 1 Constructional detail of a DC link converter

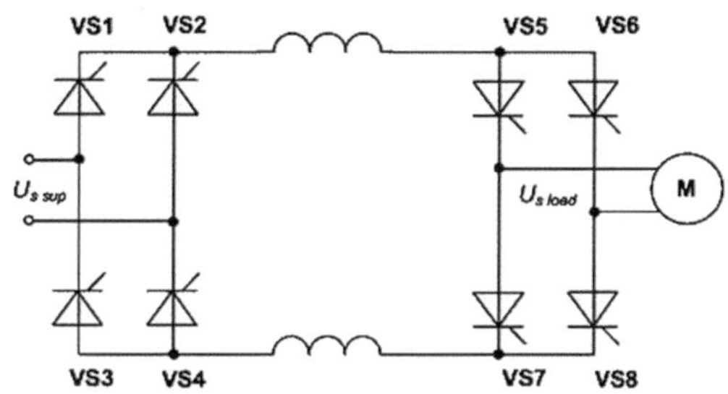

Fig. 2 Circuit diagram of a single-phase cycloconverter

2.1 Disadvantages of DC Link Converter

(1) There could be a disorder in the normal operation of other equipments due to non-sinusoidal characteristics of line current.
(2) It can also disturb the clients whose voltage can be fluctuated due to harmonics.
(3) It cannot be used in systems with fast operating characteristics.

3 Cycloconverters

They are naturally commutated direct frequency converters. They are used in large power applications up to 10 MW. Cycloconverters can be of two types:

(1) Direct single-step cycloconverter,
(2) Indirect two-step cycloconverter.

A single-phase cycloconverter is shown in Fig. 2.

On left-hand side, the converter applied is called the supply converter, whereas on the right-hand side is the load converter, due to that it can work both in rectifier as well as inverter mode. If we vary the firing angle in each half cycle which is achieved by varying the reference with time, then the harmonic content of the output is reduced. It has two types of converter in which one is known as negative converter and the other which carries positive current is called positive converter. This cycloconverter consists of two sub-converters, one with positive load current known as positive converter and the other as negative load current known as negative cycloconverter. As the operation in the entire four quadrants is possible, so it does not matter the load is active or passive. We should consider the system to be symmetrical for constant net power. The input power is solely responsible for the output power in a cycloconverter as there is no storage device present in a cycloconverter. But due to delayed firing, there is a reactive power on the line side.

3.1 Disadvantages of Cycloconverter

(1) Cycloconverters are preferred for installation above 20 kVA, as it consists of large number of thyristors which is not justified for large installation [3].
(2) For reasonable power output and efficiency, the output frequency is limited to integral multiple of the input frequency.
(3) At low output voltage, the power factor is low.

4 Matrix Converter

Basically, the application of power electronics is to convert electric energy to different forms. For this, the research is going on and many semiconductor devices and power electronic converters are invented. One of these is matrix converter which provides a single-step AC–AC conversion. The word matrix converter was firstly given in the papers of Venturini and Alesina in 1976 [4]. This paper proposed a method known as Venturini method which is also the basis of my work. In this method for fixed frequency and fixed amplitude input voltages, the gate drive signal for nine bidirectional switches is calculated for variable frequency and amplitude. Matrix converter is being studied since last 25 years. Though this power converter is still not installed in industries due to its bidirectional switches, it has synchronization and the protection problem [4]. Some attractive features of matrix converter are mentioned below:

(1) It consists of nine bidirectional switches.
(2) It works on a range of frequency and also generates 3-Φ multilevel voltages.
(3) The output is obtained in sinusoidal form, and to work on sinusoidal wave is easier.
(4) It has the property of regeneration of energy (Fig. 3).

The building block of a matrix converter is the bidirectional switches. The modulation strategy used is high frequency waveform synthesis technique. That is why it is important to study about the configuration of bidirectional switches [5].

The bidirectional switches in matrix converter need a voltage blocking capability in both directions also capable of carrying current of both polarities. But any such

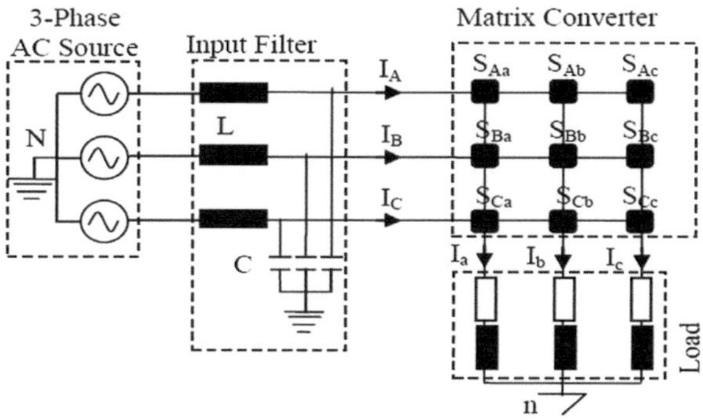

Fig. 3 Circuit of a 3 × 3 matrix converter

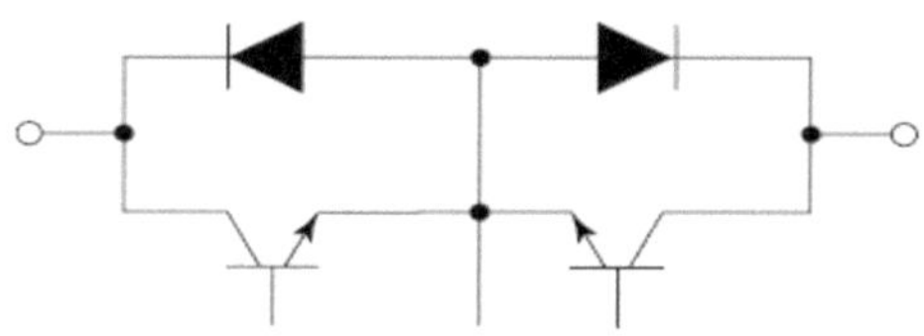

Fig. 4 Bidirectional switches

switches are not available commercially. Therefore, the bidirectional switches are constructed distinct devices. There are two constructional ways, which are mentioned below:

(i) A transistor embedded in a diode bridge arrangement,
(ii) Connection of two anti-parallel transistor and two anti-parallel diodes (Fig. 4).

5 Result

The analysis was performed by system simulation tool named as 'Simulink' which is basically a software package. This software is found suitable for modeling and simulation of power electronics converters and drives. Here, MATLAB is used to calculate the switching angle which is used to control the electronic converter (Fig. 5).

The results show that the designed controllers guarantee a good response to step changes in the references. However, this basic move toward will not always create good results; depending on the current of stator and rotor and the speed of the machine it will have some effect on the final results. Here, the above waveform of Fig. 6 shows the output with harmonics components, whereas Fig. 7 is the output waveform without harmonics that is the filtered output. Figure 8 shows the comparison between the input and output frequencies of a matrix converter through Simulink.

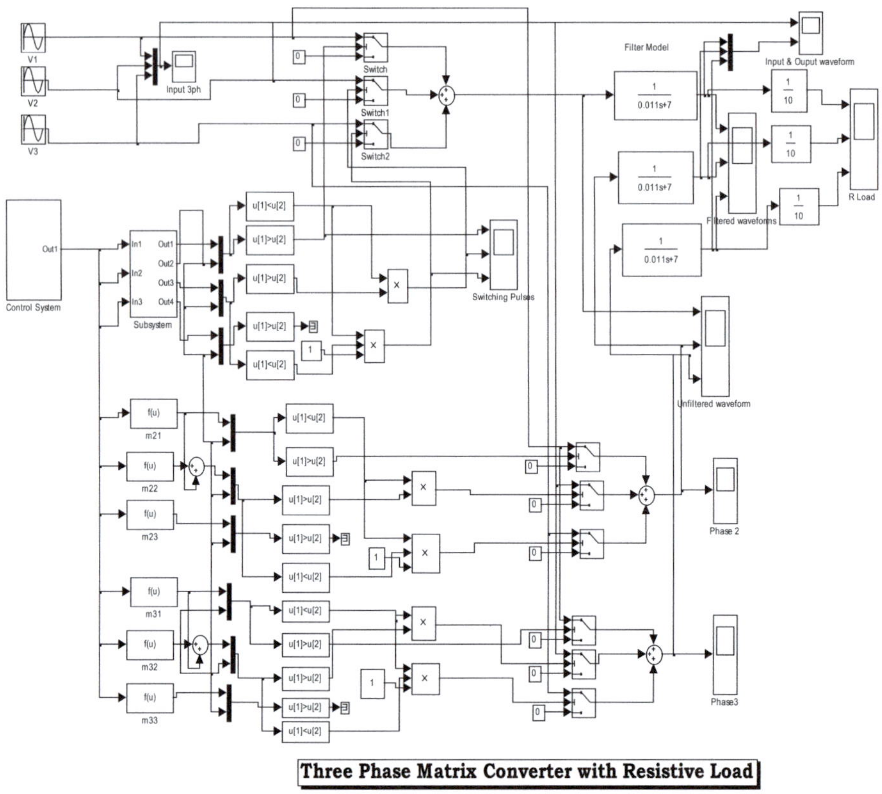

Fig. 5 Simulation circuit of matrix converter

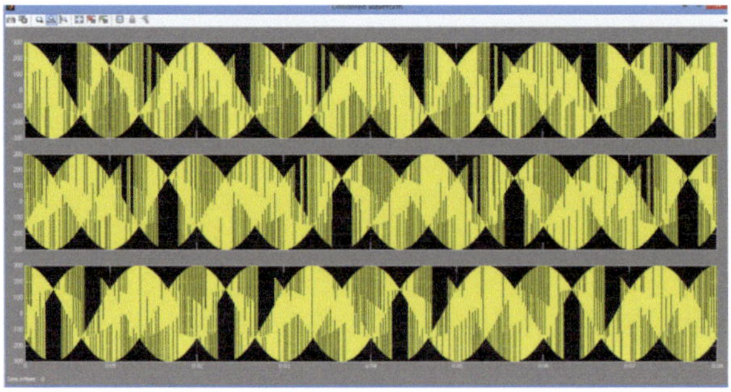

Fig. 6 Output voltage waveform at 100 Hz

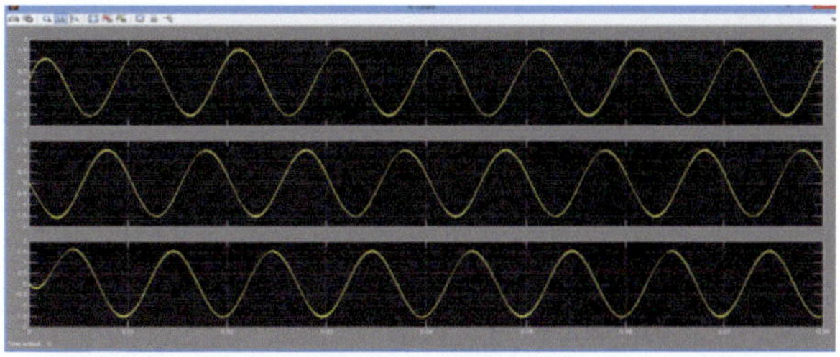

Fig. 7 Filtered output voltage waveform at 100 Hz

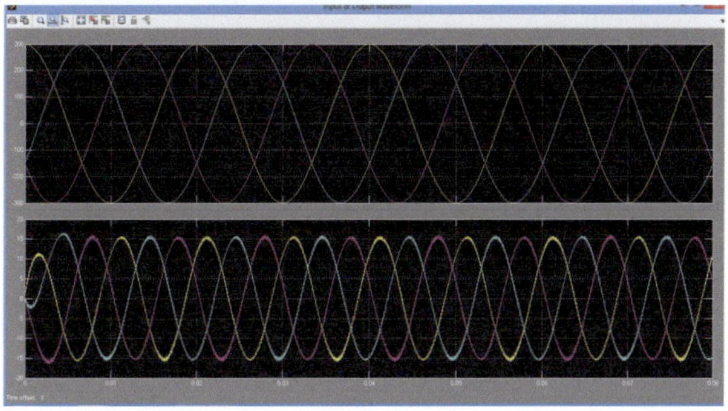

Fig. 8 Output and input voltage waveforms with input ($f = 50$ Hz) and output ($f = 100$ Hz)

6 Conclusion

The devices which convert frequency of AC at input side to another frequency at output side are known as frequency changer. A cycloconverter is the most famous and mostly used in industrial application for AC to AC transformation by switching of the device. They do not need energy storage device in their intermediate circuit. With many advantages, they had an unforgettable disadvantage also which includes its low frequency output, it cannot be greater than 0.4 times of the supply frequency, and also it has a low power factor. These problems could be removed by matrix converter with increasing efficiency as its output approaches its input. It has many advantages like bidirectional power flow, sinusoidal input and output current, and input power factor can also be adjusted easily. It is used in high temperature and serious weight applications because of elevated integration and reliability of semiconductors. Its application is limited due to the fact that its controlling technique as well as implementation is quite difficult than other converters. High level of harmonics affects the performance of equipments connected to power supply. Due to DC link, the converter becomes heavy while the efficiency is reduced.

References

1. Casadei, D., Serra, G., Tani, A., Zarri, L.: Matrix converter modulation strategies: a new general approach based on space-vector representation of the switch state. IEEE Trans. Ind. Electron. **49**(2), 370–381 (2002)
2. Kim, J.S., Sal, S.K.: Seoul National University Conference on New Control Scheme for AC-DC-AC Converters. Department of Electrical Engineering
3. Kazerani, M.: A Direct AC/AC Converter Based on Current-Source Converter Modules. IEEE Department of Electrical & Computer Engineering, University of Waterloo, Waterloo, Ontario, Canada
4. Venturini, M.G.B., Alesina, A.: Solid state power conversion: a Fourier analysis approach to generalized transformer synthesis. IEEE Trans. Circ. Syst. **CAS-28**, 319–330 (1981)
5. Popov, V.I., Zinoviev, G.S.: Matrix converter: A review of researchers in former Soviet-Soyuz and Russia. Novosiboic State Technical University Russia

Power Quality Improvement and Analysis Using Multi-pulse Converter

Renu Kumari, M. A. Ansari and Shubham Shukla

Abstract The paper displays the power quality improvement with the help of the multi-pulse converter. Proposed AC–DC converter structure, consists of phase-shifting zigzag transformer overcomes the harmonic problem in AC mains. As we know the very fact that power quality becomes a vital issue to electricity shoppers additionally to work out the ability of wattage offer devices, this work stresses on simulation to analyze the entire harmonic distortion and improve power quality during this work. Results are obtained by simulating and digital implementation of MATLAB/Simulink.

Keywords AC/DC converter · DC drives · DC motor
Multi-pulse converter · THD

1 Introduction

The power quality (PQ) issue is outlined as "any power downside demonstrates in voltage, current, and frequency deviations that lead to failure or disoperation of client instrumentation." Nowadays, power quality could be a main issue for the assembly, distribution of the electricity provide the foremost common issues, are placed like harmonics, short-term voltage deviations (sags, swells, and interruptions), future voltage variations transients, unbalance, frequency variations will cause many issues to the people who need high levels of power quality for their industrial processes or home usages [1]. The earliest harmonic distortion problems were related to third harmonic currents made by saturated iron in machines and transformers, alleged magnetic attraction masses. By definition, harmonic (or nonlinear) masses are those

R. Kumari (✉) · M. A. Ansari · S. Shukla
Department of Electrical Engineering, Gautam Buddha University, Greater Noida, India
e-mail: renu.lions@gmail.com

M. A. Ansari
e-mail: mahmadiitr@gmail.com

S. Shukla
e-mail: er.shubham.gbu@gmail.com

S. N. Singh et al. (eds.), *Advances in Energy and Power Systems*, Lecture Notes in Electrical Engineering 508, https://doi.org/10.1007/978-981-13-0662-4_18

devices that naturally manufacture a non-sinusoidal current once energized by a curved voltage supply [2]. The main target of research worker is to eliminate harmonics in shift converters. Several ways are there, and also the researchers have used the ways as per the wants and suitableness. This analysis paper deals with the reduction of total harmonic distortion victimization multi-pulse AC-to-DC conversion theme. The three-phase multi-pulse AC-to-DC system applies a phase-shifting electrical device and a three-phase system [1].

2 Power Quality and Issues

The power quality is fairly a broad concept; it may be defined as supply of voltages and system so that consumers of electric power can use electric energy with help of the distribution successfully, without disturbances in power quality defined in the IEEE 100 Authoritative as the concept of powering and grounding electronic system in a way that is suitable for the operation that is apparent and compatible with the basis wiring system and other connected equipment [3]. The power quality is damaged by different factors of the electrical network. Power quality complication affects the performance of electrical service and shortens its lifetime [4].

- **Voltage Fluctuations**

A. **Voltage Sag—**A voltage sag or voltage dip could also be a brief amount of reduction in RMS voltage which could be done by a brief circuit, overload, or starting of electrical motors. Voltage sag happens once the RMS voltage decreases between 10 and ninety p.c. of nominal voltage for simple fraction cycle to at least 1 min.

B. **Voltage Swell—**Voltage Swell is defined as a rapid short-term increase in voltage typically caused by the switching off bank of capacitors and large block of loads. Voltage Interruption is the disappearance of the supply voltage on one or more phases. When the interruption is longer than 1 min.

- **Interruptions—**Interruptions are either a short-span or long-span variation. However, the term "interruption" is usually used to consult with short-duration interruption, whereas the latter is preceded by the word "sustained" to point a long duration. They are measured and represented by their length since the voltage magnitude is often but 100% of nominal (Fig. 1).

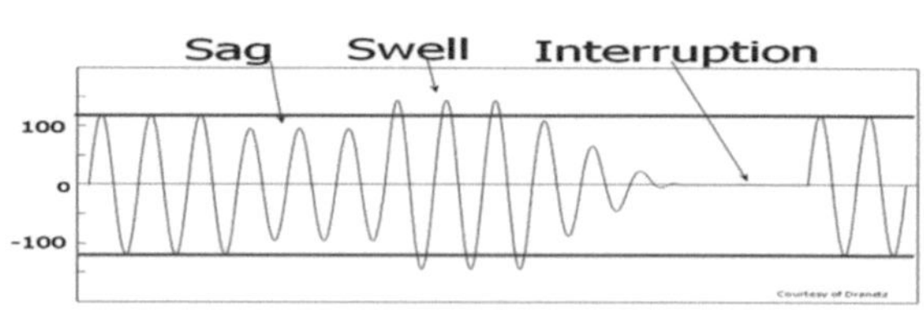

Fig. 1 Representation of power quality issues

Fig. 2 Harmonics

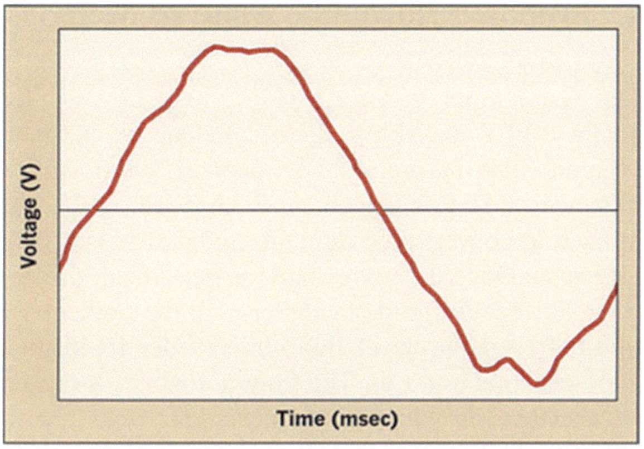

- **Harmonics**—The presence of harmonics in electrical systems suggests that current and voltage are distorted and deviate from serpentine waveforms. Harmonic voltages and currents in wattage grid give results of nonlinear electrical hundreds. Harmonic frequencies within the facility are a frequent reason for power quality problem. Harmonics in power systems lead to enhanced heating within the instrumentality and conductors, flop in variable speed drives, and torsion pulsations in motors. Reduction of harmonics is taken into account fascinating (Fig. 2).

3 Total Harmonic Distortion

Total harmonic distortion or THD may be a common measure of the extent of harmonic distortion gift in power systems. THD is outlined because of the magnitude relation of total harmonics to the worth at harmonic.

$$\mathrm{THD} = \frac{\sqrt{V_2^2} + \sqrt{V_3^2} + \sqrt{V_4^2} + \cdots}{V_1} \tag{1}$$

where V_n is the RMS voltage of nth harmonic and $n = 1$ is the frequency. Dominant sources of harmonic reproduction in power system are: controlling action of power electronic devices such as chopper, inverter causing inequality in power system leading to harmonic generation and on-linear loads like as UPS, SMPS, and battery charger [5].

4 Proposed Harmonic Analysis Methods FFT

The fast Fourier transform (FFT) may be used for analysis due to its process potency. FFT may be used to calculate the harmonic distortion and to separate even/odd/inter-harmonics. For analyzing harmonics, connected to power quality disturbances, STFTs engagement is done. FFT–FFT-quick Fourier rework analysis is utilized to convert time-domain modulation into their frequency counter parts and viceversa. Once the wave shape is periodical, the Fourier series may be helpful to calculate the magnitudes and phases of the elemental and its harmonic elements. The FFT utilizes some clever algorithms to identify identical factors. However DFT takes a lot less time to do so. The finiteor distinct is diffentiated by Fourier transform of advanced vector with n components [5].

5 Multi-pulse Method

Multi-pulse strategies involve multiple devices connected in order that the harmonics developed by one converter are off by harmonics made by alternative converters. Certain harmonics associated with some range of converters are eliminated from the power supply. In multi-pulse converters, reduction of AC input line current harmonics is incredibly vital as regards the impact the device has on the facility system. Multi-pulse strategy area unit is characterized by the utilization of multiple converters or multiple semiconductor devices with a typical load. Usually, diode is used with a better range of pulses for reducing harmonics in AC mains and drop price of ripple voltage within the DC output. These are developed in 6, 12, 18, 24, 30, 36, 48 pulses, etc. [6] (Fig. 3).

6 Phase Shifting and Harmonics

The best way to eliminate harmonics is to use a method referred to as "phase shifting." The thought of part shifting involves separating the supply into many outputs, where every output being part shifted whereas the opposite outputs with an applicable angle are eliminated. The concept is to displace the harmonic currents so as to bring them to a 180° part shift, so they cancel one another out [7] (Fig. 4).

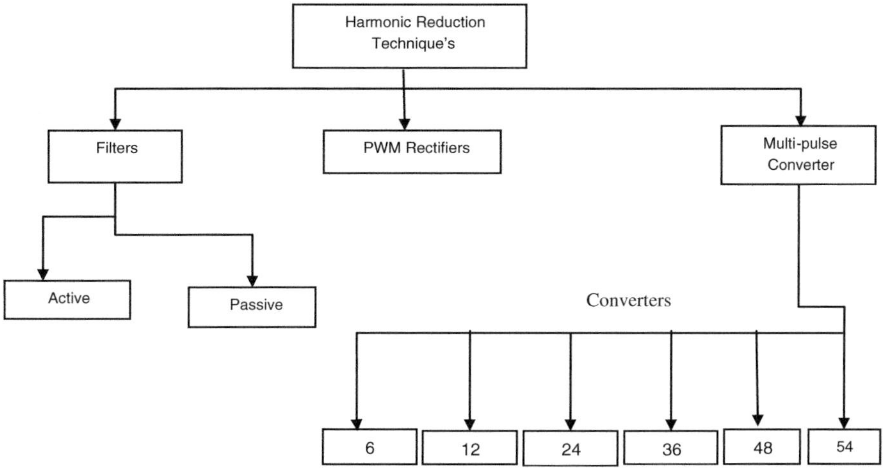

Fig. 3 Block diagram of harmonic reduction techniques

Fig. 4 Zigzag winding configuration

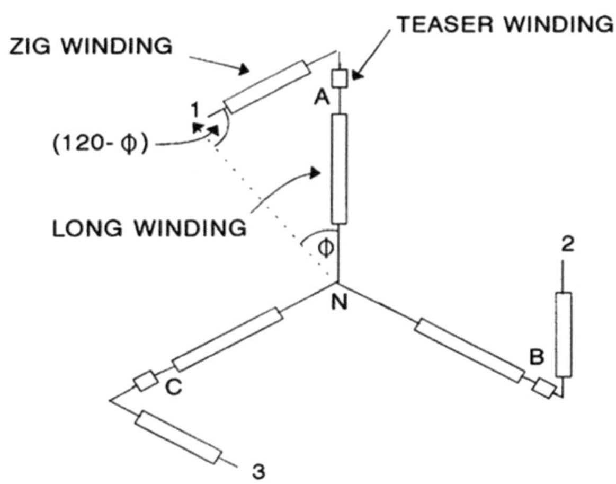

7 Simulation and Results

With the assistance of multi-pulse methodology, the harmonics contents in AC–DC converter is reduced due to multiple converters utilized in circuits so that harmonics made by one converter are off by harmonics made by another converter. There are two forms of pulse converter consistent with dominant techniques: First one is uncontrolled multi-pulse converter, and other is controlled multi-pulse converter. Moreover, an "Uncontrolled multi-pulse has mounted output and controlled converter has controlled output which can be controlled with the firing angle "α" know that just in case multi-pulse converter because the range of pulse increase the harmonics contents can

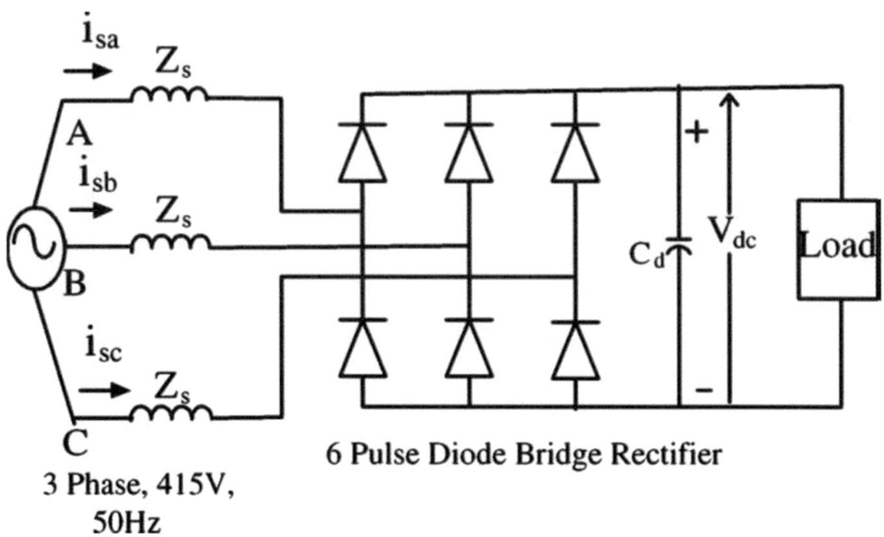

Fig. 5 Six-pulse diode bridge rectifier

be reduced and therefore the THD is relied upon part shift of the multi-pulse converter and it is given by equation . Phase shift = 60°/number of 6-pulse converter (Fig. 5).

8 Simulation of Uncontrolled Multi-pulse Converters

1. Uncontrolled 6-Pulse Converter

Uncontrolled 6-pulse device is the elemental device unit of HVDC transmission issued for rectification wherever power flows from the AC aspect to the DC aspect and inversion wherever the power flow is from the DC aspect to the AC aspect. Thyristor valves operate as switches that activate and conduct current once dismissed on receiving a gate pulse and are forward-biased. The characteristic AC aspect current harmonics generated by 6-pulse converters are $6n \pm 1$. For uncontrolled model, we used diode (Figs. 6, 7 and 8).

2. Uncontrolled 12-Pulse Simulink Model

Twelve-pulse converter could be a series association of two totally controlled 6-pulse converter bridges and needs two three-phase systems that are spaced excluding one another by thirty electrical degrees (Figs. 9, 10 and 11).

3. Uncontrolled 18-Pulse Simulink Model

Uncontrolled 18-pulse Simulink model during this 18-pulse topology is the circuit worked as same as the 6-pulse converter; thus, this topology is comparatively the

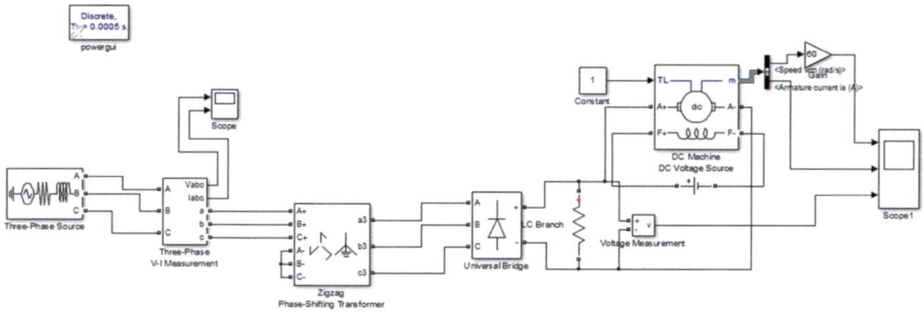

Fig. 6 Simulink/model of 6-pulse uncontrolled converter

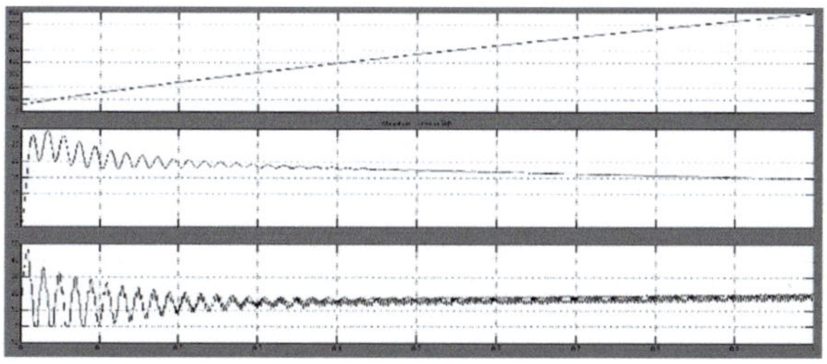

Fig. 7 Scope of 6-pulse uncontrolled converter

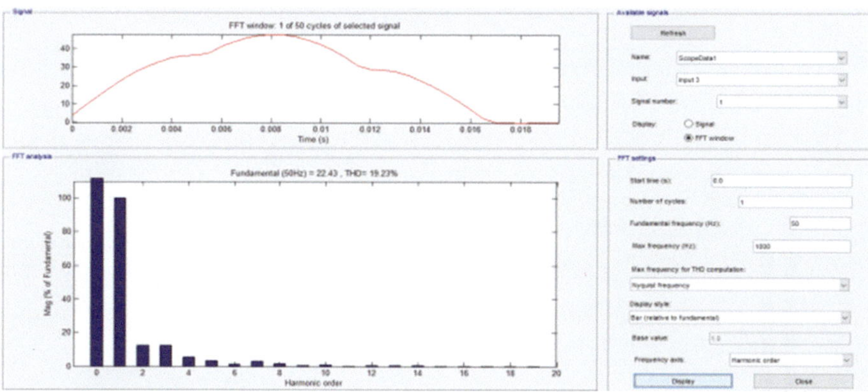

Fig. 8 THD of output voltage

hottest one. A part shift of two hundred has been provided between all three-phase shift transformers with star-connected secondary (Figs. 12, 13 and 14).

4. Controlled 6-Pulse Simulink Model

For the simulation of controlled converters in place of the diode bridge, we have a tendency to use the thyristor bridge and additionally the corresponding pulses are given (Figs. 15, 16 and 17).

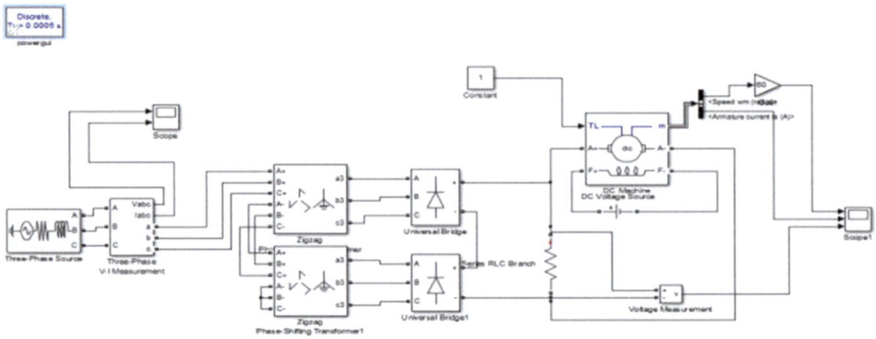

Fig. 9 MATLAB/Simulink model of 12-pulse uncontrolled converter

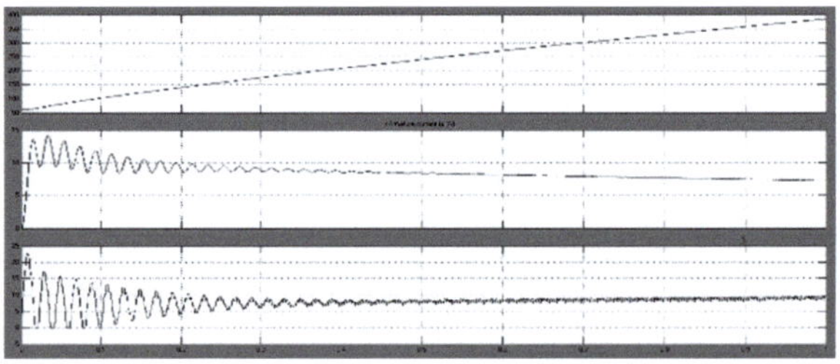

Fig. 10 Scope of 12-pulse uncontrolled converter

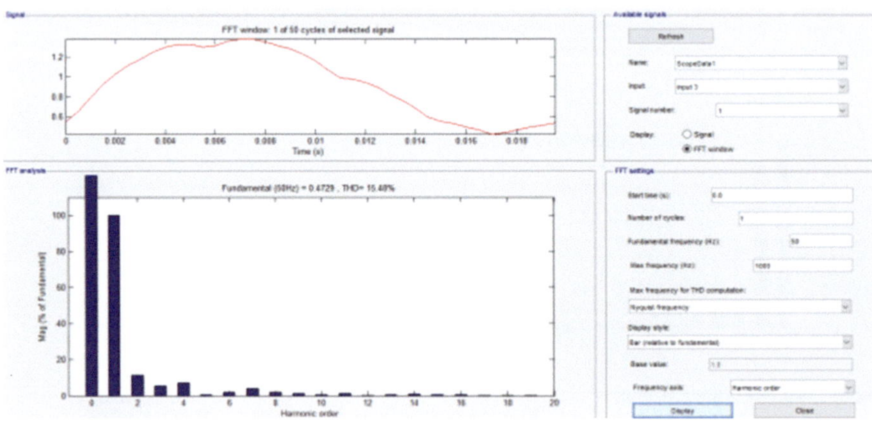

Fig. 11 Vab (line voltage) and its THD

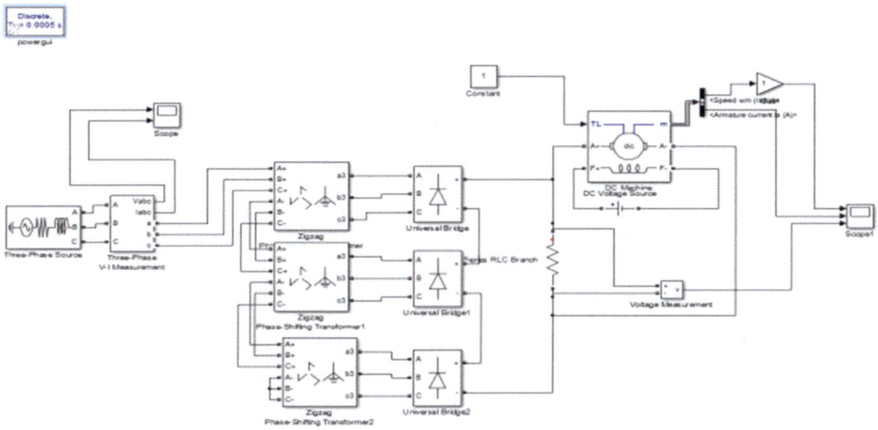

Fig. 12 Simulink/model of 18-pulse uncontrolled zigzag phase-shifting transformer

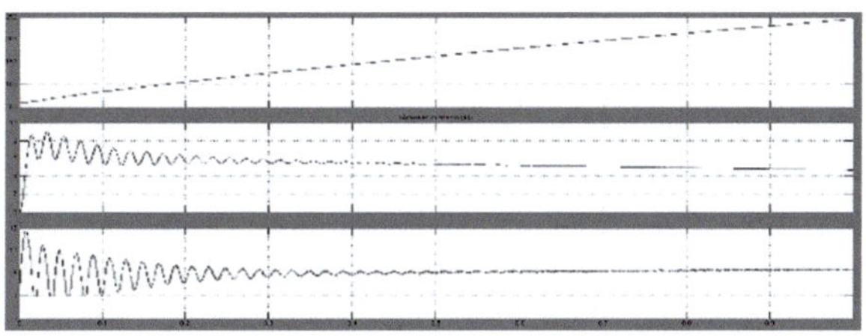

Fig. 13 Scope of 18-pulse uncontrolled converter

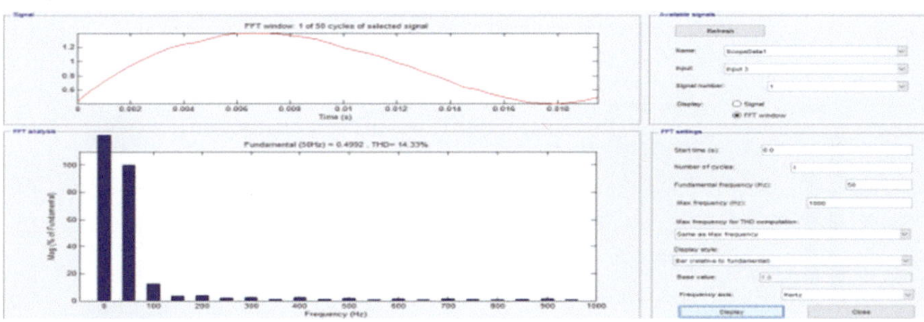

Fig. 14 Vab (line voltage) and its THD

5. Controlled 12-Pulse Simulink Model

Simulink model of 12-pulse device has two six-pulse converters connected serial or parallel in keeping with the appliance. The 3-section input of those two six-pulse converters has thirty section distinctions to every alternative. Due to this, 5th and 7th harmonics content are canceling to every alternative and 11th and 13th are assertive harmonics content (Figs. 18, 19 and 20).

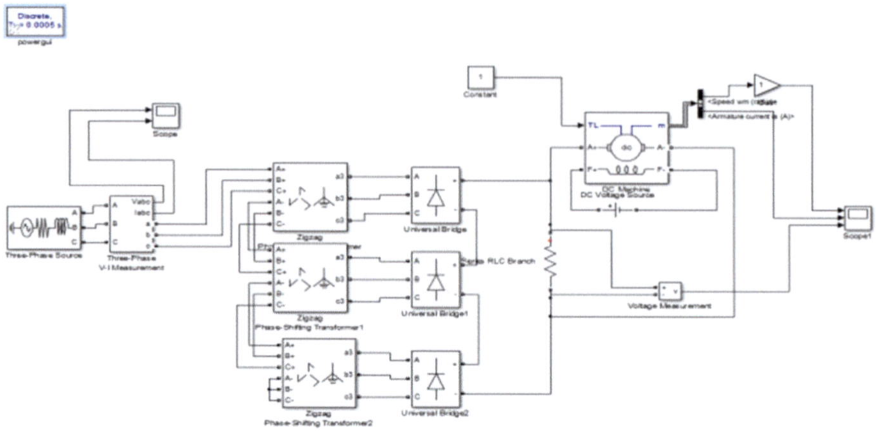

Fig. 15 Simulink/model of 6-pulse controlled converter

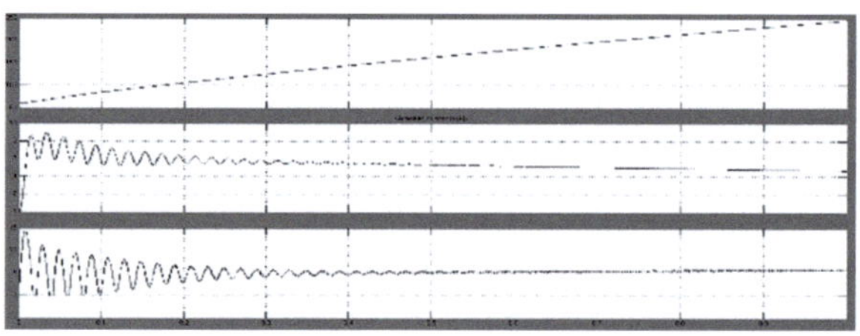

Fig. 16 Scope of 6-pulse controlled converter

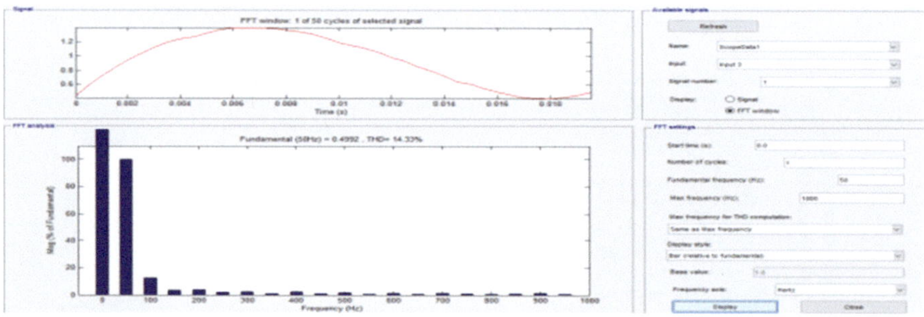

Fig. 17 Vab and its THD

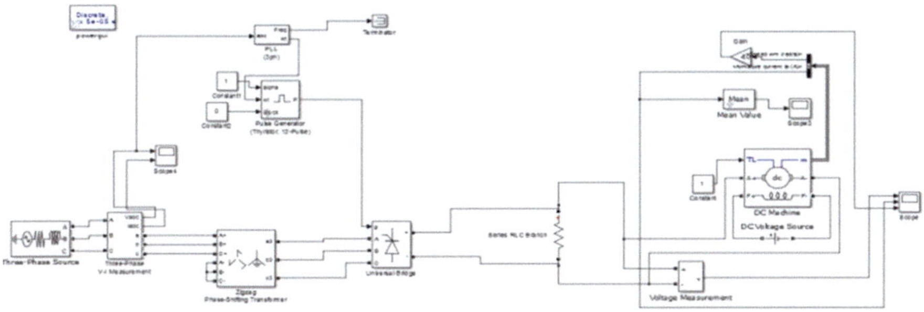

Fig. 18 Simulink/model of 12-pulse controlled converter

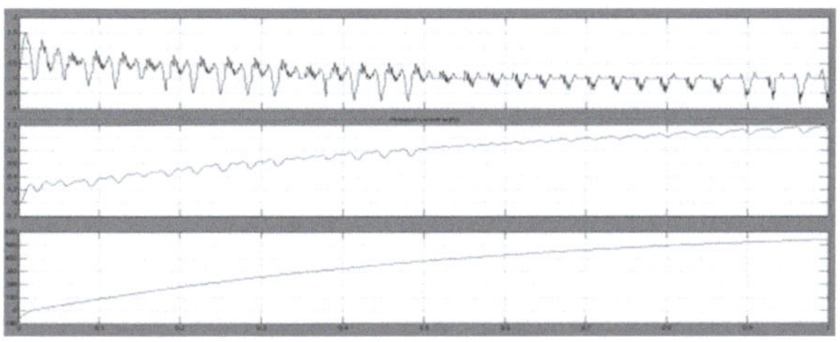

Fig. 19 Scope of 12-pulse controlled converter

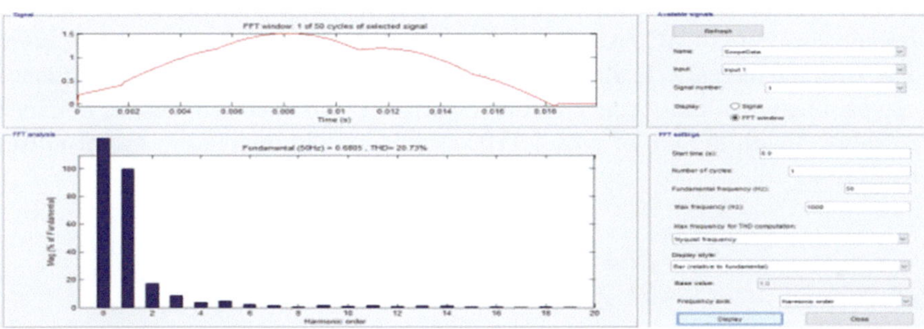

Fig. 20 Vab and its THD

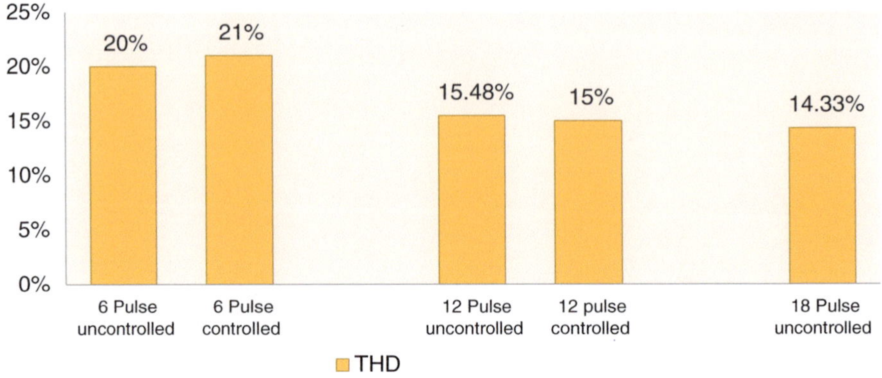

Fig. 21 Performance comparison of multi-pulse converters

9 Conclusion

In this paper, a performance comparison of multi-pulse converter has been administered. Six-pulse uncontrolled, 12-pulse uncontrolled, 18-pulse uncontrolled, and conjointly 6-pulse, twelve-pulse controlled AC/DC converters have been designed and simulated in MATLAB/Simulink. These converters are calculated in terms of harmonic spectrum of AC providing current, total harmonic distortion. Thus, commonly with the increases in varieties of pulses in multi-pulse case, the performance of the parameters of those converters is remarkably improved. The results obtained are analyzed with the no. of pulse; THD decreases with increasing no. of pulses (Fig. 21).

References

1. Tariq, M., Iqbal, M.T.: Power quality improvement by using multi-pulse AC-DC converters for DC drives: modeling, simulation and its digital implementation. J. Electr. Syst. Inf. Technol. **1**(3), 255–265 (2014)
2. Tabatabaei, N.M., Abedi, M., Boushehri, N.S., Jafari, A.: Multi-pulse AC-DC converter for harmonic reduction. Int. J. Tech. Phys. Probl. Eng. **6**(1), 210–219 (2014)
3. Malikarjuna, A., Ravindra, S., Mahendra Kumar, S.: Improving the power quality by reducing the harmonics in DC drives. Int. J. Eng. Res. Technol. **1**(1) (2012). 2278–0181
4. Patel, M.A., Patel, A.R., Vyas, D.R., Patel, K.M.: Use of PWM techniques for power quality improvement. Int. J. Recent Trends Eng. **1**(4) (2009)
5. Rejish, R.: Determination of deviation measure for power quality signals. In: International Conference on Energy Efficient Technologies for Sustainability (ICEETS), pp. 516–520 (2016)
6. Singh, D., Mahala, H., Kaur, P.: Modeling & simulation of multi-pulse converters for harmonic reduction. Int. J. Adv. Comput. Res. **2**(5) (2012)
7. Gawre, S.K., Patidar, N.P., Nema, R.K.: Application of wavelet transform in power quality: a review. Int. J. Comput. Appl. **39**(18) (2012)

MIX
Papier aus verantwortungsvollen Quellen
Paper from responsible sources
FSC® C105338

If you have any concerns about our products,
you can contact us on
ProductSafety@springernature.com

In case Publisher is established outside the EU,
the EU authorized representative is:
Springer Nature Customer Service Center GmbH
Europaplatz 3, 69115 Heidelberg, Germany

Printed by Libri Plureos GmbH
in Hamburg, Germany